An Introduction to Engineering

What it takes to make it

April K. Andreas, Ph.D.
Engineering & Mathematics

Bernard Smith, Ph.D.
Physics

McLennan Community College

About Your Textbook

This is a specialized textbook intended to help the beginning college student manage the transition to college engineering. Whether you are fresh out of high school or have been in industry for years, entering an engineering program can be a bit crazy. Mathematics, engineering, and science professors are notorious for throwing equations on the board, and then, staring at a bleak-looking classroom full of terrified students, stamping on the floor, getting red in the face, and yelling, "You should already know this!" Therefore, another major goal of this book is to help fill in any gaps in your knowledge so that when you do take your first college level courses in your major, you will be able to focus on learning the "new" material, since you'll already have all the basics down solid.

This textbook contains material from multiple authors, including material prepared by Dr. April Andreas and Dr. Bernard Smith, of McLennan Community College. It is our hope that this book will serve as a valuable reference throughout your academic career as you pursue an engineering degree. All comments and feedback about this textbook are welcome.

All material contained herein is © 2012. Permission to reproduce in whole or in part must be obtained in writing from the authors.

Becoming an Engineer

Getting Acquainted

What is engineering? Engineering is essentially made up of people who are presented with rather complicated real-world problems and try to find economical and effective solutions.

What makes a good engineer? A good engineer…

- … has a "solid working knowledge" of mathematics, chemistry, physics, programming, human factors, manufacturability, reliability, and anything else that comes up along the way.
- … may have absolutely no idea how to solve a problem when he or she starts looking at it, but eventually will figure it out, whether by research, experimentation, analysis, collaboration, or any combination thereof.

What makes a good engineering student? A good engineering student…

- … genuinely enjoys math.
- … has a passion for figuring things out on multiple levels.
- … is willing to do homework on weeknights, weekends, over Thanksgiving and Spring Break, and any other time it's required.
- … will at some point in his or her academic career feel like the dumbest person in the universe.
- … will at some point in his or her academic career feel like the smartest person in the universe.
- … will spend 5 hours looking at a problem that has a ten-minute solution.
- … knows that no matter how much work it takes, the professional fulfillment and financial rewards will ultimately far outweigh just a few more sleepless nights.

Types of Engineering

To become an engineer, you're going to have to study. You're also going to have to pick a major. There are three main branches of engineering. Forgive the crude definitions, and know that they are in no way all-encompassing, and that these fields can overlap. People actually spend lifetimes arguing these definitions.

- **Civil** – deal with things that allow society to function on a basic level (think: roads, water, power distribution, buildings)
- **Mechanical** – deal with things that make work or life easier (think: land/air/sea/space vehicles, industrial equipment, hardware)
- **Electrical** – deal with the aspect of things that require power to function (think: communication systems, radar, electronics)

Most universities will have at least one or two of these three fields of study, in addition to any number of other specialty areas. These might include agricultural, petroleum, industrial, systems, chemical, construction, manufacturing, mining, or nuclear, for example.

The best way to get information about any career, as well as updated data about job growth, salaries, and similar items is to visit the US Bureau of Labor and Statistics at http://www.bls.gov. Search for whatever field you think you may be interested in, and go from there.

Another important thing to do is to actually look at the classes you'll be taking with that major. Find a university offering a major you think you might be interested in and look at the course requirements. If the thought of taking "Soil Mechanics" makes you want to jump off a bridge, then Civil Engineering most likely isn't for you. The same thing goes for "Thermal Systems Design" and Mechanical Engineering, or "Signals and Systems" for Electrical Engineering.

Typical Classes for a Civil Engineer

- Introduction to Environmental & Civil Engineering
- Statics
- Mechanics of Deformable Bodies
- Mechanics of Materials
- Water Resources Engineering
- Structural Analysis
- Structural Design
- Environmental & Civil Engineering Design I & II
- Soil Mechanics & Foundations
- Environmental Engineering Principles & Processes
- Introduction to CAD
- Transportation Engineering & Traffic Planning

Typical Classes for a Mechanical Engineer

- Introduction to Mechanical Engineering
- Information Technology & Society
- Statics
- Dynamics
- Thermodynamics + Lab
- Mechanics of Deformable Bodies & Solid Mechanics + Lab
- Fluid Mechanics + Lab
- Heat & Mass Transfer + Lab
- Engineering Materials
- Manufacturing Processes
- Thermal Systems Design
- Design & Control of Mechanical Systems + Lab
- Elements of Machine Design
- Mechanical Engineering Design I & II
- Vibrations

Typical Classes for an Electrical Engineer

- Fundamentals of Electrical Engineering
- Electronic Circuits I + Lab
- Circuits Analysis I
- Design & Analysis of Signals & Systems + Lab
- Digital Computer Logic + Lab
- Solid State Devices
- Electronic Circuits II + Lab
- Electromagnetic Field Waves
- Statistical Methods in EE
- Introduction to Digital Signal Processing
- Microprocessors + Lab

Some Things to Know Before You Get Too Far Into This

You cannot get a two-year engineering degree. Engineering is a four-year degree, *assuming you start out in your fall semester in Calculus I, take 17 hours a semester and never drop or fail a course.* Realistically, it's closer to a five-year degree plan, particularly if you include work semesters. If you're starting off in College Algebra, plan on five or six years to completion.

If you hate math, this is not the field for you. One feature that differentiates the field of *engineering* from *engineering technology* is that an engineer will typically do a great deal of mathematical and physical analysis on a design throughout its life cycle. If you like putting things together, but aren't that interested in calculating stress, factors of safety, or signal degradation, you may consider going into engineering technology instead.

Getting Real-World Experience in College

There are many excellent ways to get experience as an engineer while you are still studying. The good news is that most of them pay, and pay pretty well. If you can, get at least two (or even more!) of these experiences on your resume, and you will find yourself quite competitive in the job market.

Internships

These are pretty standard – you come in to a company for a semester (usually the summer), work at whatever they ask you to do, collect your paycheck, and you're done. It's a good way to try out an industry or company once, see if you like it, and then either move on or ask to come back again next year. This is probably the most commonly known but actually a not-so-commonly taken path to getting engineering experience.

Co-ops

These are the ultimate in professional engineering experience. A co-op (short for "cooperative") program is where a university says, "We'll allow you to leave school on and off for a few semesters so you can get some good experience." On the other side of that coin, a company will agree to hire its students for multiple semesters.

Here's how it works: You go to school like a regular student for the first two years. After that, you begin alternating semesters going to school and going to work.

- Fall Year 3 – School
- *Spring Year 3 – Co-op (take one evening/online class)*
- Summer Year 3 – School
- *Fall Year 4 - Co-op (take one evening/online class)*
- Spring Year 4 – School
- *Summer Year 4 – Co-op (take one evening/online class)*
- Fall Year 5 – School
- *Spring Year 5 – Balance Co-op and School as needed to finish graduation requirements*

This is a fantastic agreement all-around.

It's great for the school because the companies will provide on-the-job training that the universities couldn't. It also improves their job placement rate, a statistic that measures how many of their students get jobs immediately after graduation, which boosts their national rankings.

It's great for the companies because they get cheaper labor (often paying between $12 and $20 an hour) and they get to "test you out" before actually hiring you on full-time. Also, if you're a good student and you enjoy your job, then theoretically you also have other smart friends that you'll recruit to their company. Viola! They've stolen all the top students from the competition before they even had a chance to tempt you!

Finally, it's great for you because you'll get experience, you'll get paid, and often you'll get a job offer upon graduation. Often, one co-op semester will pay for your whole next semester at school, thereby reducing the amount of loans you may otherwise need to take out. Some schools will even give you credit toward your degree for completing a certain number of co-op terms.

Also, some companies will begin treating you as a "real employee" starting with your first co-op term, including perks like 401(k) matching or length-of-service credit for pensions and retirement. Companies may also have social activities for co-op students, and sometimes pay for housing or relocation.

And of course, the best scenario is that if you do a fantastic job with your co-op, that you just might get an offer when you graduate!

To find a co-op, you need to find companies that have co-op programs. You also have to be going to a university that has a co-op program. (Be wary of one that doesn't.) Some universities will have a full-time staff member in charge of helping to place students with companies, and sometimes you'll have to find the job and apply for it on your own.

REUs

A Research Experience for Undergraduates (REU) is kind of like an internship but for academia. Rather than spending a semester working for a company, you'll spend a semester (usually a summer) working for a university doing research. If you think you might want to go to graduate school, this is an excellent option.

REUs usually pay pretty well, but instead of paying hourly, you'll get a lump sum "stipend" for the entire period of time. REUs will often pay for housing or arrange for you to stay on campus for reduced rates. If you're lucky you might get to be a co-author on a paper in a scientific journal, or get to work with a world-renown expert on a topic that you are passionate about.

Applying for REUs is pretty straight-forward: find one that you're interested in online, complete any required paperwork, sending transcripts or letters of recommendation as appropriate, and waiting for a response.

Other Smart Ways to Earn Cash

Aside from the avenues mentioned above, there are ways to earn money during the semester while still going to school full time.

Tutoring – This is the bread and butter of engineers. While you're struggling with partial differential equations, there are others out there struggling with the FOIL method. This is where you can make some cash. You'll be able to help freshmen figure out vectors and titration and sophomores understand the shell method and mesh analysis. To find tutoring jobs, advertise on your own, ask at your school's learning center, or find a local tutoring agency. Tutoring is flexible – you can usually make your own hours. Not only is tutoring a great way to earn money, teaching something to someone is the best way to really learn it.

Lab Assistant – Did you really enjoy your mechanics lab? Did programming just "make sense" to you? Was chem lab a breeze? If so, go talk to your instructor for that class and see if he or she is looking for some help. Getting hired as a lab assistant might mean that you walk around helping students during the lab, or that you grade the lab notebooks, or both. Either way, it's a fun, on-campus job that will help to keep you from forgetting all that good stuff that you learned!

Creating Your Engineering Resume

Getting Started

Remember everything valuable, relevant, and interesting that you've ever done or learned in your life. Now fit that all on one page. This is the joy of resume writing.

Putting together a resume can be a long and painful process. The good news is that if you create your resume once and do it well, for the rest of your career the only real work you will have to do is update it every time you need it.

In this chapter, we will focus specifically on putting together a resume for an internship, co-op, or research position search. As you gain professional experience and are out of college for a few years, the rules of resume-writing start to change.

Your resume should be divided into sections, usually along these lines:

- Personal Information (name, address, email, phone number)
- Objective
- Education
- Relevant Work Experience
- Other Skills and Interests

You may also include sections such as:

- Other Certifications and Experience
- Honors and Awards
- Research
- Publications
- Conferences

You may choose to combine or otherwise reorganize these categories based on your own experience. You should always start with your strongest points, and then work down to the less strong or less relevant ones. The example resume in this chapter shows how this may be done.

General Tips

- Fit your entire resume on one page. Unless you have over 10 years of relevant technical work experience, you need to highlight just the key facts.
- Always list the most recent work, education, publication, anything first. Resumes are always in *reverse-chronological* order. The oldest stuff goes last.
- Take your resume to your campus writing or career center for advice.
- If you are applying for a job at a defense contractor or government agency, you should indicate on your resume your citizenship or residency status, since that can impact whether or not that company is allowed to hire you.

Personal Information

The first thing anyone should see on your resume is your name. It should be big, bold, and centered. The next thing on your resume needs to be how to get in touch with you. Include a permanent mailing address, a phone number that can receive messages, and an email address that you check regularly.

Some notes:

- If you are constantly moving between apartments, friends' couches, and other transitory living environments, it can be hard for potential employers to get in touch with you. Try to find a family member or trusted friend with a permanent address who is willing to accept mail on your behalf and list that address instead.
- Whatever phone number is on a resume, make sure your voicemail message is professional. You don't want an interviewer's call to go to voicemail only to hear, *"Hey, guys, it's me. I'm probably passed out somewhere after blowing off class and gaming all night, so leave me a message and maybe I'll call you later if I feel like it."* Go with a more traditional, *"This is _____. Please leave a message,"* or some variation on that theme.
- Your email address can speak volumes about your maturity. The best email addresses are firstname.lastname@plainolboringdomainname.com. In general, your email address should indicate nothing about you except how to get in touch with you, so addresses like seniors2011rock@hotmail.com, dist12champs@aol.com, or the ever-so-awful amfii99300039@oldhighschooldomain.com. If you don't want to change email addresses, sign up for a new account with a professional name, and forward that account to the one you check all the time. It's probably better not to even use your college or university email account in case your resume is discovered after you are no longer a student.

Objective

Your objective statement is where you give the potential employer an idea of the kind of job you are looking for. The statement should be general enough that you are seen as a potential candidate for a large number of jobs, yet not so general that you come across as vague or wishy-washy.

Your objective should always be to *contribute to a company in some way*, not just to get experience or get to do something fun and interesting. The idea is that you want to serve your employer, not just get something out of the company without giving back. Your objective should also be revised for each company to which you are applying, or even each job within the same company.

Education

If you are new to the professional workforce, probably the most relevant thing you have done is go to school. This should therefore be listed immediately after your objective statement. List your school and the type of degree awarded (or seeking). If you are in a community college, state that you are pursuing an Associate of Science, for example, and that you anticipate majoring in fill-in-the-blank engineering. If you have graduated recently, you may choose to indicate the year, but it's not required. If you have not yet graduated, you should list your anticipated graduation year.

If you are right out of high school and are in your first two years of college, go ahead and list your high school under education, but by the time you graduate from college, you can drop that. By the time you have completed over 60 hours of technical coursework, your college education is much more important, so in the interest of conserving space (you only have one page, remember?) just drop the high school stuff.

If your GPA is a 3.0 or better, go ahead and list that. If your GPA is a 3.5 or better, definitely list it. Many companies have GPA minimums to even consider interviewing a candidate, so if yours is not listed, you may not get a call. If your GPA is below a 3.0, do whatever is necessary to get it above that threshold value. Engineers with GPAs less than 3.0 can have a really hard time finding a job.

Sometimes your engineering GPA may actually be higher than your overall GPA. If English and Government are dragging you down, list your engineering GPA separately. You may also choose to calculate an "Engineering, Math, and Science" or similar GPA.

As a reminder, for schools with an A, B, C, D, F system:

Grade	Grade Points
A	4.0
B	3.0
C	2.0
D	1.0
F	0.0

To calculate your GPA, multiply the number of credit hours a course was worth (often the second number in a four-digit numbering system) times the appropriate number of grade points. Then add those up and divide by the total number of hours completed. That's your GPA. Don't include courses you haven't completed.

Course	Credit Hours	Grade	Grade Points Earned
ENGR 1201	2	4.0	8
ENGR 1304	3	3.0	9
MATH 1314	3	4.0	12
MATH 1316	3	3.0	9
CHEM 1411	4	2.0	8
CHEM 1412	4	3.0	12
Total Credit Hours	**19**	**Total Grade Pts Earned**	**58**

Eng, Math, Sci GPA = 58/19 = 3.1

As you complete coursework, it can help to list those courses directly on your resume. You should probably organize your coursework by subject matter. This way, an employer can say, "Okay, this person has had fluid mechanics – great, that's just what we need," or,

"Fantastic – this person has had a few programming courses." By listing courses you make it easier for someone to know exactly what you can and cannot do.

Relevant Work Experience

When listing your work or volunteer experience, first list the company you worked for, your job title, and the dates worked there. Use short, bulleted points describing what your responsibilities were. Be specific. If you did something out of the ordinary or had responsibilities beyond what are normally expected for this kind of work, be sure and say so.

This is an engineering resume. Although you may have spent several years as a waiter/waitress or folding t-shirts at a clothing store, that kind of work experience is not going to impress someone hiring for an engineering position. Although you may choose to keep some of these kinds of jobs on your resume, think about other technical experience you may have had (paid or unpaid).

You may have had experience tutoring, working as a computer tech, or mentoring schoolchildren with science fair projects. These are all valid experiences to put on a resume. (If you need to, retitle this section, "**Relevant Work and Volunteer Experience**.") If you've had work doing construction or electrical work, welding, or plumbing, definitely put this on there. Some of the best engineers out there are ones with a strong background in the more "hands-on" aspects of their field.

If you have to list nontechnical work, show how that job gave you valuable experience in team leadership or team building, time management, recruiting, or sales. Really think about what you learned in that job, and try to go beyond the surface. If you did spend hours cataloging 20-pound bags of aquarium gravel, were you also in charge of reordering and maintaining a minimum level of stock? That's responsibility. Were you a host or hostess at a restaurant in charge of putting people at different tables while juggling waitstaff requirements and complaints? Those are people skills.

Highlight your own contribution. Use words or phrases like:

- I organized ...
- I planned ...
- I led a team which designed...

Now that you've had to go through the process of putting together an engineering resume with non-technical experience, let that be an incentive for you to go out and find some more technically-oriented work. Here are some options that you're probably not going to find in "Help Wanted" or on Craigslist:

- Ask a professor if he or she needs help in a lab, or if someone needs a TA for grading.
- Ask at the tutoring center if they need help, or employ yourself as a tutor.
- Volunteer at a church or community center to help with science-based educational programming.
- Volunteer as a computer tech at a local school for the summer.

One thing you must be is willing to work for little or nothing until you have the skills that warrant that big paycheck. The more technical experience you get, the more employable you will become.

Other Skills and Interests

This section can be a catch-all for other kinds of things you'd like a potential employer to know about. An interviewer may sometimes use these "Other Interests" to break the ice at the beginning of the interview and to get to know you better.

Whatever interests you have, they should be respectable. Clearly, don't list things like "Voted Most Likely to Do Anything" or "Winner of South Park Sing-a-long Contest 2006 – 2009." This is a section where you can put that you were a team captain of your high school football team or were in the chess club. If you were an Eagle Scout or Valedictorian, you should certainly list that here. If you speak several languages, that would be good to include as well.

Dos and Don'ts

Don't:

- Don't title your resume, "Resume". Title it with your name.
- Don't label your name, as in "Name: Arkanis Gath." The same goes for all your other personal information. It's perfectly obvious what your name is.
- Don't use the same resume for every job application.
- Don't forget to include important non-traditional experience you may have that shows the quality of your character or work ethic.

- Don't email a resume to a recruiter in a file titled, "Awsum Rsme" or anything similar. Instead, change the filename to your first and last name, followed by the word "Resume," like "Arkanis Gath Resume."

Do:

- Keep electronic copies of your old resumes. You never know when you'll want to highlight an old job or research experience.
- Customize your resume for each job that you are applying for.
- Spell-check your resume.
- Be honest. If you try to embellish your experience or qualifications, that will become obvious in the interview, and you will look very bad.
- Convert your resume to a pdf before submitting it to an employer. Use a free service like www.pdfonline.com or download something like pdfcreator from sourceforge. PDFs are much more professional and you won't have to worry that a recruiter won't be able to read your filetype.
- Complete some additional research into writing resumes on your own. An example resume is included below, but there are many more great examples available at the end of a simple Google search.

Sample Resumes

On the next two pages, you will find example resumes for people with very different backgrounds. (The borders around the resumes are cosmetic for the purposes of emphasizing the pages in this book; you should not put a border around your actual resume.)

The first resume is for Arkanis Gath, a recent high school graduate with a bit of experience working construction. Arkanis still lists his high school graduation on his resume since it is relatively recent and his resume highlights education over experience.

The second resume is for Liia Jannath, who has been out of high school for a while and is returning to college. Liia already has an Associate's, which she lists, and her resume highlights her real-world experience. Although her technical experience is limited to her volunteer work, her resume emphasizes a strong work ethic as well as management and sales skills. She also explains an apparent "demotion" at her current job, showing that she made a commitment to return to school and consciously chose to reduce her work duties.

ARKANIS GATH

1234 Rocket Road
Waco, TX 77000

arkanis.gath@gmail.com
(xxx) xxx-xxxx

OBJECTIVE
I am seeking a co-op position at Space Exploration Technologies where I can leverage my education and experience to contribute as an engineer.

EDUCATION
McLennan Community College
Associate of Science (expected completion May 2015)
With focus in mechanical engineering and mathematics
Cumulative GPA: ~3.73
Math/Science/Engineering GPA: ~3.91

Midway Independent School District
Midway High School – focus on math, music
Graduated May 2010 – GPA: 3.96

RELATED COURSE WORK AND SKILLS

McLennan Community College
Engineering and Science:
Introduction to Engineering, Programming for Engineers*, General Chemistry I with Lab, General Chemistry II with Lab*, General Physics I with Lab*
Mathematics:
College Algebra, Calculus I, Calculus II*
* In progress

Texas State Technical College
Practical Application:
Intro to Surveying, Plane Surveying, Technical Drafting

Other Skills Include: C++, Matlab, HTML, Microsoft Excel, Microsoft Word, Microsoft PowerPoint

RELEVANT WORK AND VOLUNTEER EXPERIENCE

Math Tutor, McLennan Community College
August 2012 – present
- Explain developmental math and college algebra concepts
- Teach students to use computer to solve math problems

Lakeshore Community Center
January 2012 – present
- Serve as a volunteer to coordinate monthly events
- Responsible for selecting daily activities for after-school programs

B&R Construction Corp.
May 2010 – June 2011
- Supported teams in residential and industrial construction
- Responsible for small crew, including scheduling and materials
- Assistant to lead carpenter

HONORS, AWARDS AND ACTIVITIES

Eagle Scout – Boy Scouts of America
Honor Roll – McLennan Community College
Varsity Letter, Team Captain – Midway High School Football

Liia Jannath

735 Falcon Drive
Waco, TX 77000

jannath@yahoo.com
(xxx) xxx-xxxx

OBJECTIVE
I am seeking a co-op position at Space Exploration Technologies where I can leverage my education and experience to contribute as an engineer.

EDUCATION
McLennan Community College
Associate of Science (expected completion May 2015)
With focus in electrical engineering and computer science
Cumulative GPA: ~3.52
Math/Science/Engineering GPA: ~3.85

Associate of Arts with a Field of Study in Business (Completed 2002)

RELATED COURSE WORK AND SKILLS
McLennan Community College
College Algebra, Trigonometry, Calculus I*, Introduction to Engineering, Engineering Graphics*, General Chemistry I with Lab*, Fundamentals of Programming I and II*
* In progress

Other Skills Include: Java, Microsoft Office, fluent in Spanish

WORK AND VOLUNTEER EXPERIENCE

Heart of Texas Builder Supply
Senior Sales Associate, 2011 – present
- Provide quality service on the sales floor
- Help contractors manage orders and payment plans

Assistant Manager, 2002 – 2011
- Managed a crew of 20 employees
- Responsible for setting schedules and resolving employee conflicts
- Managed inventory and set competitive prices
- Increased sales volume an average of 15% per year
- Requested a reduction in duties in order to return to school

Sales Associate, 2000 – 2002
- Worked the sales floor in multiple departments
- Specialized in working with local contractors
- Top sales rep based on customer surveys, 2000 – 2002

BEST Robotics, 2010 – present
- Serve as coach for Robinson High School design team
- Provide suggestions for students in design of robot
- Help troubleshoot design, particularly software problems

Sales Representative, Barters & Sons Insurance, 1999 – 2000

COLLEGE AWARDS AND ACTIVITIES

Phi Theta Kappa
President, Engineering and Physics Club

Tartan Scholar
Honors College

Surviving Your Interview

Getting Through Engineering Interviews

Your resume is perfect. You got the interview. You're going in tomorrow. Now what? Advice abounds online for how to survive the interview. We will cover some of the basics here and warn you of some of the major traps.

The Night Before the Interview

Make sure you've done your research on the company. What are their major projects/products? Look them up in Google News and see what recent articles have been posted about them. If you've had any friends who either work for the company or have interviewed with them, call them and see if they can offer any additional insight or advice.

Try to drive to the interview location beforehand so you make sure you know where you're going and how long it takes to get there.

Make sure you have collected your "Interview Preparedness Packet," including:

- Several copies of your resume
- Directions to the interview location
- The interviewer's contact information
- An unofficial copy of your transcript(s)
- A tidy-looking folder or portfolio to keep all your papers in
- Any additional information the interviewer has asked you to bring

Other things you need to make sure you have taken care of:

- Have all your clothes (including belt, nice socks/hose, and shoes) laid out, pressed, and clean where the cat won't lay on them. Your outfit should cover your shoulders, knees, and ankles. Women can get away with bare elbows, but men should have sleeves all the way down to the wrist. If you have tattoos, piercings, or other such accoutrement, keep it under wraps. Make a statement *after* you get hired.
- Prepare some questions to ask your interviewer. It is common to begin closing up an interview with, "Do you have any questions for me?" Saying, "Uh, no, not really," will not impress anyone.
- Prepare your "elevator speech." The other common way to end an interview is to say, "Is there anything else you want to mention before we're done here?" This is

where you need to have a 30 – 45 second speech prepared where you tell the interviewer how excited you are to be there, highlight any special skills that you want to make sure they know about, and thank them for interviewing you. Again, "Uh, no, not really," is not the way to go.
- Make sure your cell phone is plugged in and charging.
- Make sure your car is full of gas.
- Set your alarm clock. Set another alarm on your phone. Call a friend or family member and ask them to call you as well to make sure you wake up.

The Morning of the Interview

Shower, eat, and make sure you've got your interview packet and phone before you head out the door. Drive safely and listen to something on the radio that will get your brain ready for this interview. Get yourself in "The Zone" and be ready and alert. Once you get to where you're going and you've checked in, turn off your cell phone. (Not vibrate – off. You don't want to be in the middle of a discussion with an interviewer and jump out of your chair after an unexpected buzz!)

During the Interview

This is Go Time. You've done all the preparation you can, and now it's all up to you. You need to be your best. Here are some good tips:

- If there is more than one interviewer, don't assume that the oldest person is the only one you need to impress. Show equal respect to everyone in the room. Be sure and shake everyone's hand.
- Sit comfortably, but not too comfortably. Pretend you're taking someone out for Valentine's Day dinner at an expensive restaurant, not having friends over to watch the Super Bowl.
- For most questions, avoid yes or no answers. You should always elaborate.
- Smile. Visual cues make up a lot of how an interviewer will see you.
- Make it seem like all you have ever wanted to do in life since you were a little baby was to work for fill-in-the-blank company doing fill-in-the-blank work. Interviewers should come away with a sense of your enthusiasm for working for *their company*, not just your enthusiasm for having a job in general.

There are some times when you may be asked an inappropriate question. These could include:

- Do you have any kids?
- What church do you go to?
- When did you graduate from high school?
- How old are you?
- Are you married?
- Do you plan on having kids soon?
- What's your native language?
- Do you smoke or drink?

These questions aren't just Nobody's Business. They're also illegal and you don't have to answer them. Most of the time when these questions are asked it is an "honest mistake." The interviewers may just be friendly and curious. However, sometimes they can have a more serious undertone. Whatever the case, your religion, marital status, age, number of children, etc., is a personal and private matter.

The best way out of these questions is to say something like, "I'm not really sure how that's relevant," and leave it at that.

Companies can ask if you are over 18, if you speak specific languages, or if you are authorized to work in the U.S. They can ask if you can meet a specific work schedule or if you are willing to work overtime or travel, but they cannot ask what religious holidays you observe. They can ask if you take illegal drugs, but cannot ask about prescription or over-the-counter medication in general.

That said, there are some jobs where you may need to be eligible for a security clearance. An interviewer may say something along the lines of, "This job requires U.S. citizenship and the ability to obtain a security clearance. Is there anything in your background that would prevent you from meeting these requirements? I only need a yes or no answer, and you don't need to provide any details."

Behavioral Interviewing

The current trend for interviewing is for interviewers to focus on how you have handled situations in the past, as a means to judge how you would handle them in the future. Most of the questions are very broad, and you should be prepared for them. Practice your answers in advance. If you don't think you have an answer to a question, think harder. Somewhere at some point, you have almost assuredly run into these kinds of situations.

Common questions run along the lines of, "Can you give me an example of a time when…"

- … you had to make a split-second decision.
- … you had to deal with a difficult boss.
- … you disagreed with a course of action.
- … you had to make an unpopular decision.
- … you made a bad mistake.
- … you led a team to achieve a goal.
- … you worked under time pressure.
- … you failed at something.
- … you saw an unethical situation unfold.
- … you achieved a long-term goal.

Your response shouldn't be just to describe the incident and why it difficult to navigate; instead, your response should focus on the *solution*. How did you wade through this tough experience and come out a better and more learned person? There are many websites available that discuss behavioral interviewing in detail.

The Technical Interview

For some positions, it is not uncommon to be given a "test" as part of the interview process. Did your resume say that you can code a PHP email server? Go for it.

The only way to be prepared for these tests is to study hard in school, learn everything well (and not do a brain dump after the final), and not claim to be able to do something that you cannot do.

Traps to Avoid

There are some sure-fire ways to ruin an interview. Here are some examples:

Lying on your resume. If your resume says that you spent last summer creating a telecommunications network for under-privileged youths in Cambodia, you had better have a good story to go along with it. If you claim that you can program in LISP, don't be surprised if your interviewer asks you to sit in front of a computer and code up something simple.

Thinking you can fake it. If you are a junior-level engineering student, your interviewer is going to assume you know something about forces and vectors. If you have only gotten by in your engineering classes off the "kindness" of others, it's going to show. Thinking that someone's going to hire you just because you have "people skills" but no significant technical background is going to find you in a sorry state of unemployment. If you are going in for a highly-competitive job, it's not a bad idea to make sure you remember the basics, like what Newton's laws are, just in case someone asks.

Not practicing. Most colleges and universities have career or writing centers that will not only help you with your resume, but also help with mock interviews. These services are usually free, and a trained professional will walk you through a typical interview, and then go over the results with you afterward. If you can't find a professional, ask a friend to pretend to be an interviewer. Come up with some questions you think an interviewer might ask and practice answering these questions until the answers come out smoothly and comfortably.

Assuming the Interviewer Knows Everything About You Already. Did you work on a complicated project that took significant effort? Is it only given one line on your resume as "Research Assistant"? Make sure that when you answer questions during an interview that you provide specific details that show off your knowledge and experience. This is not the time to be modest.

Acting Like You Already Got the Job. Don't ask about benefits. Don't ask about salary. If you want to know about benefits and salary, ask the Human Resources person after you get the offer. Usually the person interviewing you has no idea what kind of salary you would be getting. Don't ask whether or not it's okay that you take two weeks off in July to go to Acapulco. This will likely accomplish nothing but to make the interviewer think, "Man, I haven't even hired this person yet, and already they're asking for special treatment!" Be humble, honest, and friendly.

Acting "Me-Centered." As you answer questions, tell about your own accomplishments, but also be sure to acknowledge helpful teammates and mentors. Show how you're a team player. Don't ask questions of your interviewer like, "Will you send me to get training with LabView?" or, "Will I get to work on cool projects?" Instead ask, "Is there any training that you would recommend that I get before I would start in this position?" or, "What kinds of projects is the company working on now where I may be able to contribute?" Your interviewer should come away feeling like you are excited about *serving the company*, not the other way around.

Not being honest about your own weaknesses. Many interviewers still ask, "What are your greatest strengths and weaknesses?" Be ready for it. If you had a difficult experience with a team, don't make your answer seem as though you are whining about how nobody understands you. If you've made mistakes in the past, admit them, and show how you've learned from these mistakes to become a better worker.

Making the Interviewer Assume Responsibility for Your Successful Interview. The interviewer should not feel like they are pulling teeth trying to get an answer out of you. If you feel like the interviewer is doing more talking than you are, they may be compensating for you not doing any of the talking. Make it easy on them by being well-prepared, comfortable, and confident.

Sample Behavioral Interviewing Worksheet

On the next two pages is a sample worksheet an interviewer might use. Practice going through this process as many times as you can before your interview. Remember that luck is where preparation meets opportunity, and the better prepared you are, the better chance you have of landing that dream job!

Sample Behavioral Interviewing Worksheet

Icebreaker Questions

Question	Candidate Response	Interviewer Comment
Why do you want to work for us?	*Remember to emphasize here why you want to work for that company <u>in particular</u>, not just why you want a job in general.*	
How far are you along in school? When do you think you'll graduate?	*Be confident – make it clear that you have set a timeline for yourself and you're going to meet those goals.*	
Interviewer choice (GPA, something related to resume or field of study)	*The interviewer is still trying to help you feel comfortable at this point – try to take advantage of this early opportunity to show off certain skills or knowledge.*	

Behavioral Interviewing Questions (Pick three of 10)

Question Asked (1 – 10)	Candidate Response	Interviewer Comment
	Be sure and avoid yes/no answers, and take every chance you can to brag on yourself. If you don't say it, the interviewer doesn't know it. *When you tell stories about things that happen, make sure your story has a clear and concise conclusion, with you coming out on top of the situation.*	

Question Asked (1 – 10)	Candidate Response	Interviewer Comment
	By the second of these three questions, make sure that you're really selling yourself. You should be emphasizing characteristics about yourself that would make someone want to hire you as an engineer.	
	You're getting close to your last chance to land the job. Make sure you're giving all the relevant details – don't assume that the interviewer knows the intricacies of the situation you are describing.	

Potential Questions

1. Tell me about a time when you had to make a split-second decision.
2. Tell me about a time when you had to deal with a difficult boss.
3. Tell me about a time when you disagreed with a course of action.
4. Tell me about a time when you had to make an unpopular decision.
5. Tell me about a time when you made a bad mistake.
6. Tell me about a time when you led a team to achieve a goal.
7. Tell me about a time when you worked under time pressure.
8. Tell me about a time when you failed at something.
9. Tell me about a time when you saw an unethical situation unfold.
10. Tell me about a time when you achieved a long-term goal.

Follow-up Questions

Question	Candidate Response	Interviewer Comment
Do you have any questions for me?	*When in doubt, ask about the interviewer's own experience in the company.*	
Is there anything else you want to mention before we're done here?	*If you're too brain dead to think at this point, a simple handshake and, "I just really appreciate the opportunity to come out here and speak with you today," will suffice.*	

Professional Development Plans

Let's look at goal-setting, or in the professional world, "Professional Development Plans," or PDPs. PDPs are all the rage in industry and academia, where you list your professional goals from year to year, and your strategies for achieving them. You will often be asked to update your PDP semiannually or quarterly, depending on your organization's policies, and this update is often accompanied by a meeting with your supervisor to go over your progress.

Often, PDPs are coupled with a self-evaluation. The idea is that you evaluate yourself, and then set goals to improve areas where you may fall short. Note that it is never okay to rank yourself as Perfect across all categories, even if you really think you are. No one else is going to believe you, and they'll assume it means that you don't know how to evaluate yourself.

Note also that while setting personal goals (exercise, diet, spending more time with family) is an excellent activity, these kinds of goals should not appear on your PDP. The PDP should only have "professional" goals, like learning a new analysis tool or becoming more involved in professional organizations.

Here are some tips for creating a PDP and self-assessment:

- Your boss may be looking at this PDP five years from now when deciding on whether or not to give you a promotion or raise. Make sure it looks professional.
- All your goals should be specific and measurable. ("Work harder" is not specific. "Complete all homework assignments on time" is much better.)
- Make sure everything is formatted the way your boss wants it.
- Make sure everything is spelled correctly.
- Your comments should make it seem like you spent more than 3 minutes working on this. They should be well thought-out and insightful.

Since you're still in school, creating an "academic PDP" is a fantastic exercise. An example of a PDP for an engineering student follows on the next two pages, followed again by how an updated PDP would look a month later.

Arkanis Gath
Self-Assessment and Professional Development Plan
Fall 2013, September 1

Academic Skills Self-Assessment

Use this scale:	Strongly Disagree	Disagree	Neutral	Agree	Strongly Agree
	1	2	3	4	5

Skill	Sep 1	Oct 1	Nov 1	Dec 1
I have clear academic goals and I know what I want.	5			
I allocate enough of my time to do well in school.	4			
I seek out my professors to get help on material that I don't understand.	2			
I don't skip assignments, and turn in all required material on time.	3			
I go to every class, pay attention, and take good notes.	3			
I study for exams using my notes, the textbook, and any other resources I can find, including YouTube and other websites.	4			
I always master the material presented in previous classes before the next class period.	2			
I work with other students, taking advantage of study groups, without "copying" or "cheating."	4			
I am confident in myself and my ability to succeed in engineering.	5			
I spend a large amount of time on campus, studying, talking with other people in my classes, visiting professors, and otherwise taking advantages of the resources available.	5			
I would give myself an A+ on the quality and amount of time and energy I devote to my education.	4			

Professional Development Goals

Goals	Strategy	Outcome
1. Complete every homework assignment in Cal 1 on time.	1. I'm going to make a checklist with all the homework assignments and their deadlines. I'll put that checklist in the front of my binder and each time I complete an assignment, I'll check it off and write in the grade earned.	1. TBD

Professional Development Plans

Goals	Strategy	Outcome
2. Be prepared for every Intro to Engineering Daily Activity.	2. I'm going to take careful notes every day during class in my spiral so the notes don't get out of order. If I get confused, I'll rewatch the recorded lectures and fix my notes as necessary.	2. TBD
3. Don't be late for any class.	3. I'm going to plan to get to my first class of the day 15 minutes early so that I won't be late. I will also find a friend who is in the same class and we'll text each other 30 minutes before class starts to make sure we're both up.	3. TBD
4. Read all the assignments in my English class.	4. Each time a book is assigned, I'll divide the total number of pages by the days available. I'll make a checklist and keep track of how I'm progressing through the reading.	4. TBD
5. Meet with each professor at least once during the first two weeks of class.	5. I will either find a question from the notes or homework or just stop by to say hi. First, I'll check to see if I need an appointment.	5. TBD
6. Become involved in a student activity group.	6. I'll find a group that I'm interested in and start visiting meetings. I'll volunteer to work or attend at least one event/activity within the first semester.	6. TBD

Arkanis Gath
Self-Assessment and Professional Development Plan
Fall 2013, October 1

Academic Skills Self-Assessment

Use this scale: Strongly Disagree   Disagree   Neutral   Agree   Strongly Agree
                             1   2   3   4   5

Skill	Sep 1	Oct 1	Nov 1	Dec 1
I have clear academic goals and I know what I want.	5	5		
I allocate enough of my time to do well in school.	4	4		
I seek out my professors to get help on material that I don't understand.	2	3		
I don't skip assignments, and turn in all required material on time.	3	3		
I go to every class, pay attention, and take good notes.	3	4		
I study for exams using my notes, the textbook, and any other resources I can find, including YouTube and other websites.	4	4		
I always master the material presented in previous classes before the next class period.	2	3		
I work with other students, taking advantage of study groups, without "copying" or "cheating."	4	4		
I am confident in myself and my ability to succeed in engineering.	5	5		
I spend a large amount of time on campus, studying, talking with other people in my classes, visiting professors, and otherwise taking advantages of the resources available.	5	5		
I would give myself an A+ on the quality and amount of time and energy I devote to my education.	4	5		

Professional Development Goals

Goals	Strategy	Outcome
1. Complete every homework assignment in Cal 1 on time.	1. I'm going to make a checklist with all the homework assignments and their deadlines. I'll put that checklist in the front of my binder and each time I complete an assignment, I'll check it off and write in the grade earned.	1. I'm doing much better on this. I keep forgetting to record the grades I'm getting on assignments, but I am turning them in on time.

Professional Development Plans

Goals	Strategy	Outcome
2. Be prepared for every Intro to Engineering Daily Activity.	2. I'm going to take careful notes every day during class in my spiral so the notes don't get out of order. If I get confused, I'll rewatch the recorded lectures and fix my notes as necessary.	2. This is going great! It took a few days to figure out how to configure my computer to watch the videos, but they are very helpful.
3. Don't be late for any class.	3. I'm going to plan to get to my first class of the day 15 minutes early so that I won't be late. I will also find a friend who is in the same class and we'll text each other 30 minutes before class starts to make sure we're both up.	3. I still need to work on this. The friend that I found is missing class entirely, so I need another strategy here.
4. Read all the assignments in my English class.	4. Each time a book is assigned, I'll divide the total number of pages by the days available. I'll make a checklist and keep track of how I'm progressing through the reading.	4. I'm having a hard time keeping up in English. I just keep thinking that it won't take as long as it actually does to read, so I need to allocate more time for this.
5. Meet with each professor at least once during the first two weeks of class.	5. I will either find a question from the notes or homework or just stop by to say hi. First, I'll check to see if I need an appointment.	5. I've talked to all but one of my professors, but I'm still having trouble getting in touch with the last one.
6. Become involved in a student activity group.	6. I'll find a group that I'm interested in and start visiting meetings. I'll volunteer to work or attend at least one event/activity within the first semester.	6. I'm very happy with my progress here. I've become more involved in the Engineering & Physics club and was elected to serve as that group's Social Chair.

Professional Engineers in Texas

Introduction

Professional Engineers (PEs) are engineers who have been granted authority, usually by a government body, to sign or stamp their own drawings, reports, and calculations, and thereby assume legal responsibility for them.

Being a professional engineer can provide job security since only a small percentage of engineering graduates go on to get licensure. Being a PE carries a level of prestige, since the process to become licensed is very challenging. Also, if you ever want to be an independent consultant, being a PE gives you an extra air of professionalism and authority.

In the U.S., it is not required to get licensure to work as an engineer; however, some fields lend themselves to it more than others. Civil engineers have the highest rate of licensure, since the types of projects that civil engineers often work on, such as road or building construction, require a PE seal. If you are interested in working in other countries as an engineer, having a PE license can help smooth the way. Texas has a higher overall level of licensure than other states because many engineers work in both Texas and Mexico.

It is a common misconception that you earn a PE in a specific field, such as mechanical or electrical engineering. This is not, in fact, the case. A PE must show a general competency in general engineering knowledge and then takes a test specializing in his or her field. However, the final document simply says, "Professional Engineer." It is up to the PE to use his or her ethical and professional judgment to only work on projects that he or she is qualified to work on.

In the state of Texas, the Texas Board of Professional Engineers, or TBPE, is the agency responsible for granting Professional Engineer status. Licensure requirements may vary state, and you should plan accordingly.

The Application Process

The typical process for getting licensure in Texas is as follows:*

1. Complete an ABET-accredited engineering degree.
2. Take and pass the Fundamentals of Engineering (FE) examination, offered twice a year at locations throughout the state. There is a one-hour break between the morning and afternoon examinations.
    a. The four-hour morning examination has 120 1-point questions and is a general engineering exam that covers these topics:
        - Mathematics
        - Chemistry
        - Computers
        - Statics
        - Dynamics
        - Strength of materials
        - Material science
        - Electrical circuits
        - Thermodynamics
        - Fluid mechanics
        - Engineering economics
        - Engineering management
        - Ethics
        - Environmental engineering
    b. The four-hour afternoon examination has 60 2-point questions and the student must choose to focus on a specific area:
        - Chemical engineering
        - Civil engineering
        - Environmental engineering
        - Electrical engineering
        - Mechanical engineering
        - Industrial engineering
        - General engineering (covers the topics from the morning exam in more detail, and adds some biology)

   After passing this exam, the person becomes an "Engineer In Training" or EIT.
3. Complete four years of engineering experience, focusing on design and analysis. Document experiences and design projects as they are completed. Ideally, you should complete your work while under the supervision of a PE, since you will eventually need a PE to review and certify your work in your application.
4. After four years, prepare a Supplemental Experience Record (SER) detailing the work experience. This is a very long and laborious process and each page must be reviewed by a licensed PE before it is submitted. A candidate must also provide three references from currently-licensed PEs.

---

* This information has been simplified to some degree, so as not to bog the reader down with minutia. For other paths to licensure, including non-accredited degrees and graduate degrees, visit www.tbpe.state.tx.us for more information.

5. Take and pass the Texas Ethics of Engineering examination, which is an open-book test that can either be taken online or downloaded and completed at your convenience. When you submit this exam, you will also submit your license application and application fee.
6. The application is approved by TBPE.
7. Within two years of approval, the applicant must take and pass the Principles of Engineering (PE) examination. This exam is discipline-specific, and some disciplines are only offered once per year. Disciplines include:
   - Agricultural
   - Architectural
   - Chemical
   - Civil
   - Control Systems
   - Electrical and Computer
   - Environmental
   - Fire Protection
   - Industrial
   - Mechanical
   - Metallurgical and Materials
   - Mining and Mineral
   - Naval Architecture and Marine Engineering
   - Nuclear
   - Petroleum
   - Structural I
   - Structural II

After passing all examinations, the applicant officially becomes a "Professional Engineer." There are additional education requirements to maintain licensure.

Engineering Ethics

Why Ethics?

On Tuesday, January 28, 1986, the students of Sally K. Ride elementary school in The Woodlands, Texas crowded into the cafeteria to watch the space shuttle Challenger rocket into space. This particular launch was of particular interest to the educational community because the first teacher was going to go into space. Christa McAuliffe, the first member of the Teacher in Space Project, was chosen from 11,000 applicants and was going to hold two fifteen-minute lessons from the shuttle. It seemed like the "Ultimate Field Trip" and a perfect way to bring science into the classroom.

The launch went off and the students cheered. The camera followed the shuttle up into the sky, and 73 seconds later, there was a large explosion. Ranging in ages from six to nine, the students continued to cheer, unaware of what was going on. For the adults in the room, however, hearts stopped and hands went to cover eyes and faces. The Challenger had broken apart, disintegrating over the Atlantic Ocean and killing all seven astronauts aboard.

Immediately following the Challenger disaster, fingers pointed blame in all directions. Ultimately the technical failure was identified as a faulty O-ring, but the true failure was not in hardware but in the organizational culture at NASA, which had identified the potential flaw in 1977. The Rogers Commission, appointed by then-president Ronald Reagan, concluded that NASA experienced a massive failure in leadership and communication, and today the study of the Challenger disaster is standard reading for all serious studies of whistle-blowing and ethics.

Engineers are ultimately responsible for the life, health, and property of society, and therefore a study of ethics is crucial. Failing to report a concern, address a known safety issue, or avoid a conflict of interest could quite literally cost people their lives.

The National Society of Professional Engineers Code of Ethics

While all states have their own professional licensing organizations, there is also an organization, the National Society of Professional Engineers (NSPE) that promotes licensure and allows professional engineers to network across the country. Though each state can publish its own particular ethics code, the NSPE has established six "Fundamental Canons" of engineering ethics. The following text is taken directly from their website:*

"Engineers, in the fulfillment of their professional duties, shall:

1. Hold paramount the safety, health, and welfare of the public.
2. Perform services only in areas of their competence.
3. Issue public statements only in an objective and truthful manner.
4. Act for each employer or client as faithful agents or trustees.
5. Avoid deceptive acts.
6. Conduct themselves honorably, responsibly, ethically, and lawfully so as to enhance the honor, reputation, and usefulness of the profession."

Consider these as you read the case studies in the following chapters.

---

[*] This information and more details can be accessed at http://www.nspe.org/Ethics/CodeofEthics/index.html.

Ethics Case Study 1: Larom, Inc.

This and the other case studies presented in this book were developed through the National Science Foundation, with Grant No. DIR-8820837, headed by Michael S. Pritchard, Director, Center for the Study of Ethics in Society. These case studies are part of the public domain and are included here verbatim. Additional case studies and resources on engineering ethics can be found at http://ethics.tamu.edu/.

Larom

I

A recent graduate of Engineering Tech, Bernie Reston has been employed in the Research and Development (R&D) Chemical Engineering Division of Larom, Inc. for the past several months. Bernie was recommended to Larom as the top Engineering Tech graduate in chemical engineering.

Alex Smith, the head of Bernie's unit, showed immediate interest in Bernie's research on processes using a particular catalyst (call it B). However, until last week, his work assignments at Larom were in other areas.

A meeting of engineers in Bernie's unit is called by Alex. He announces that the unit must make a recommendation within the next two days on what catalyst should be used by Larom in processing a major product. It is clear to everyone that Alex is anticipating a brief, decisive meeting. One of the senior engineers volunteers, "We've been working on projects like this for years, and catalyst A seems to be the obvious choice." Several others immediately concur. Alex looks around the room and, hearing no further comments, says, "Well, it looks like we're in accord on this. Do we have consensus?"

So far Bernie has said nothing. He is not sure what further testing will show, but the testing he has been doing for the past week provides preliminary evidence that catalyst B may actually be best for this process. This is also in line with what his research at Engineering Tech suggested with somewhat similar processes. If catalyst B should turn out to be preferable, a great deal of money will be saved; and, in the long run, a fair amount of time will be saved as well. Should he mention his findings at this time, or should he simply defer to the senior engineers, who seem as determined as Alex to bring matters to closure?

II

Bernie somewhat hesitantly raises his hand. He briefly explains his test results and the advantages catalyst B might provide. Then he suggests that the unit might want to delay its recommendation for another two weeks so that he can conduct further tests.

Alex replies, "We don't have two weeks. We have two days." He then asks Bernie to write up the report, leaving out the preliminary data he has gathered about catalyst B. He says, "It would be nice to do some more testing, but we just don't have the time. Besides, I doubt if anything would show up in the next two weeks to change our minds. This is one of those times we have to be decisive--and we have to <u>look</u> decisive and quit beating around the bush. They're really getting impatient on this one. Anyway, we've had a lot of experience in this area."

Bernie replies that, even if the data on B is left out, the data on A is hardly conclusive. Alex replies, "Look you're a bright person. You can make the numbers look good without much difficulty – do the math backwards if you have to. Just get the report done in the next two days!"

Bernie likes working for Larom, and feels lucky to have landed such a good job right out of Engineering Tech. He is also due for a significant pay raise soon if he plays his cards right.

What do you think Bernie should do? Explain your choice.

1. Write up the report as Alex says.
2. Refuse to write up the report, saying he will have no part in falsifying a report.
3. Other.

III

[Following II. 1.] Bernie decides to write up the report. When he is finished, Alex asks him to sign it. Bernie now has second thoughts. He wonders if he should sign his name to a report that omits his preliminary research on catalyst B. Should he sign it?

IV

Bernie has now had more time to do research on catalyst B. After several weeks his research quite decisively indicates that, contrary to the expectations of Alex and the other more experienced engineers in the unit, catalyst B really would have been, far and away, the better choice. What should Bernie do now?

1. Keep the data to himself – don't make trouble.
2. Tell Alex and let him decide what, if anything, to do.
3. Other.

V

Bernie decides to say nothing. Although Larom has lost a lot of money by investing in an inferior catalyst, it is quite possible that this is the end of the matter for Bernie. The customer never complains, and no one outside at Larom raises any questions. However, it might go otherwise. Suppose a Larom competitor discovers that catalyst B is better for this type of work and it begins receiving contracts that Larom would normally be awarded. Further, what if Alex's superior then makes an inquiry into why his unit has missed out on this development?

VI

[Following II. 3.]

Bernie tries to convince Alex that a straightforward report should be submitted. Since there is a virtual consensus in the unit that catalyst A is best, A can be recommended. But the preliminary evidence about B can also be mentioned. After all, Bernie suggests, if the entire unit is convinced that A is best despite the preliminary evidence about B, why wouldn't those outside the unit be persuaded by the received wisdom of the unit? If they aren't persuaded, perhaps they will grant the unit more time to continue the research on B.

Somewhat to his surprise, Bernie finds Alex and the others receptive to his suggestion. The preliminary evidence about catalyst B is included in the report, even though A is recommended.

Unfortunately, Alex's superiors are very upset with the recommendation. They are unwilling to go ahead with the project without further testing, but they bitterly complain that the further delay will be very costly. Alex is severely criticized for not having a more convincing set of data. He, in turn, blames his staff, especially Bernie, the new specialist in this area. Bernie, Alex tells his superiors, failed to complete the necessary testing in a timely fashion. Alex tells his superiors that he should have supervised Bernie's work more closely, and he assures them that he will not let matters get out of control again. Although Bernie is not fired, he is not promoted and his salary is frozen for another year. What should he do?

1. Nothing. No good will come from complaining.
2. Confront Alex, telling him what you think of what he has done, but carrying it no further.
3. Other.

VII

Bernie decides he has nothing to gain from complaining to Alex or anyone else about becoming the "scapegoat" of the project. So, he keeps quiet. Sometime later, Alex is being considered for promotion to another division. Members of Bernie's unit are privately interviewed about his performance in the unit. Bernie is told that his comments will be kept confidential. What should he say in his interview?

VIII

Bernie says nothing negative about Alex in the interview. None of the others in the unit do either. Alex is promoted to another division. However, a year later it is discovered that he has directed someone in his new division to falsify data for reasons very similar to those in Bernie's original situation. The new person does what Alex asks. The result is a significant loss of money to Larom--only this time there is an expensive product-liability lawsuit relating to an unsafe Larom product. An inquiry takes place. The person who has falsified the report says that Alex has often requested that data be falsified--and that he typically has gotten young engineers to do the "dirty work" for him. So, it comes back to Bernie. He is asked why he didn't report Alex's orders to falsify data when the matter first came up. Bernie is accused of being partly responsible for allowing Alex to be promoted--with the resulting harm to others and loss of money and reputation to Larom.

[This case is inspired by two brief case studies presented by Roy V. Hughson and Philip M. Kohn in **Chemical Engineer**, May 5, 1980: "The Falsified Data" and "The Falsified Data Strike Back." These are two of several brief case studies that they present. You might enjoy looking at the others. They are on pp. 100-107 of that issue.]

Commentaries

Michael S. Pritchard, Department of Philosophy, Western Michigan University

Although convinced there may be reason to prefer catalyst B to A, Bernie may also be convinced that deferring to the judgment of the more experienced engineers is the best course of action – especially in this kind of situation. He may actually be persuaded that the others are probably right. His is a minority view, and he is considerably less experienced. The recommendation apparently cannot wait for further testing. Besides, Alex is Bernie's division head, and Bernie may believe that his job is to do as he is told. So, Bernie may conclude, it is best to support his colleagues' recommendation -- both from the standpoint of Larom, Inc. and his own self-interest.

However, four cautions should be noted from the outset. First, although Bernie may have a general obligation to do what he is told by his superiors, blind or unthinking obedience is not obligatory. He has no obligation to do anything illegal or unethical, regardless of which "authority" requests it. In this case, it is not at all clear that Alex's superiors at Larom would approve of his effort to falsify the report, or that they would fault Bernie for refusing to comply with Alex's request. After all, the report is for them. Why would they willingly agree to be duped – especially since approving the wrong catalyst could turn out to be very costly to Larom?

Second, Bernie should be alert to the possibility of what sociologist Irving Janis calls **groupthink** (Groupthink). This is the tendency of cohesive groups to arrive at consensus at the expense of critical thinking. Janis identifies eight "symptoms" of groupthink:

1. The illusion of group invulnerability. ("We've always been right before.")
2. Shared stereotypes. ("We/they" thinking about those outside the group who may disagree – the other as "enemy.")
3. Rationalizations.
4. Unquestioned belief in the group's inherent morality. ("We're all committed to doing the right thing.")
5. Self-censorship by individual members. (Reluctance to "rock the boat.")
6. The illusion of unanimity. (Silence taken as agreement.)
7. Direct pressure applied to ensure conformity when dissenting opinions are expressed. ("We can't wait forever.")
8. Mind-guarding. (Keeping outsiders who have dissenting views from presenting their views directly to the group – "I'll pass your concerns on to the group.")

Several of these symptoms seem to be present at the initial meeting. There is evidence that at least some of the senior members of the group share the illusion of invulnerability ("We've been working on projects like this for years..."). Rationalizations for not having done more research on catalyst B follows on the heels of this illusion. Given the shared purpose of recommending the best catalyst for the job, the members may believe in the inherent morality of the group ("We know we're on the right side"). Silence in response to Alex's final look around the room for further comments may be the result of some self-censorship (especially if Bernie fails to speak up). This, in turn, feeds the illusion of unanimity. Finally, Alex's evident desire to orchestrate the group to a quick and decisive resolution indicates a readiness to apply direct pressure to any dissenters. Given that much may be at stake for Larom in this situation, Bernie is well advised to be alert to such group dynamics, rather than simply deferring his more senior colleagues.

Third, Bernie seems to be the only one with evidence that catalyst B might be preferable, and his previous work with catalyst B has already impressed Alex. If he does not speak up, who will? It is unfortunate that Alex did not assign Bernie to work on catalyst B earlier. Perhaps sometime earlier Bernie should have made a special point of discussing with his colleagues some of his previous work with catalyst B. But why didn't Alex take the lead? It seems that an opportunity for significant research when Bernie first joined the R&D Division was lost. However, shifting responsibility to Alex for lacking foresight does not relieve Bernie of responsibility for speaking up now.

Fourth, Bernie is not only asked to suppress data about catalyst B but also to alter the other data. That is, he is asked to **lie**. Alex no doubt sees this as a lie intended to "protect the truth," since he believes that catalyst A really is best. However, as Sissela Bok convincingly argues, even lies of this sort are ethically questionable (<u>Lying: Moral Choice in Public and Private Life</u>). She points out that we have a tendency to overestimate the good that comes from lying and to underestimate the harm that comes from lying. Individually and collectively lies do much to undermine trust. Also, by deceiving others, lies often lead people to make decisions they would not make if they had more reliable information, thus undermining their autonomy. Bok concludes that we should lie only after looking carefully to see if any alternatives preferable to lying are available.

One alternative that might work is for Bernie to suggest that they include all the available data but still recommend catalyst A. Since the data has not discouraged them from recommending catalyst A, why should they fear being forthright with others? As a later scenario shows, this tactic could have unfortunate consequences for Bernie, too. But this is ethically preferable to submitting a falsified report – signed or unsigned. No option guarantees there will not be complications. So, why not do what seems right and let the "chips fall" where they will?

Should things backfire as described in the later scenario, it may be understandable that Bernie would not speak out against Alex when Alex tries to lay the blame on Bernie, but it would not be wrong for Bernie to speak out even if he ends up being demoted or losing his job. Bernie's failure to speak out in the confidential interview seems highly questionable, however. Bernie has witnessed (and been victimized by) at least three examples of Alex's poor leadership: his failure to support needed research in a timely fashion; his effort to get Bernie to falsify data; and his lying about Bernie's shortcomings. Bernie faces relatively little risk in speaking confidentially about these matters in the interview; and, as we see in one of the later scenarios, he may do much good for Larom by speaking out.

Although it might seem to Bernie throughout this case that it would be **prudent** not to "rock the boat," it is not at all clear that this would be a correct assessment on his part. There are too many ways in which things can go wrong for him to be sure what a prudent course of action would be. However, prudence and ethics are not the same, and it seems that we can be more certain about what it would be ethical for Bernie to do.

Two basic lines of thought might help Bernie sort out what is at stake ethically when he is facing the initial question of whether to falsify data. One line has already been discussed –

viz., that of thinking through the possible **consequences** of doing as Alex says, and of comparing this with other alternatives. In doing this Bernie needs to consider his basic responsibilities to Larom. (How what he does might affect Larom's customers and society generally is perhaps too indeterminate to be of much relevance here.) Although in the "heat of the moment" Bernie may find it difficult to think of little else than Alex and the others pressing for closure, his responsibilities are not exhausted by relationships to his divisional colleagues.

A second line of thought rests on the idea of **universalizability**: Whatever is right for Bernie in this situation is right for similar persons in similar circumstances. It may not be easy to determine just what should count as relevantly similar circumstances, but any serious thinking about this will conclude that Bernie's situation is hardly unique -- and this thinking will not confine itself just to engineers who are deciding whether falsify data. Bernie needs to think about the more general phenomenon of lying. Just how sweeping must his acceptance of lying be in order for him to conclude, in good faith, that falsifying data in this case is justifiable from an ethical point of view? To say that the sweep is very wide indeed is not to **predict** that doing what Alex request will result in widespread lying. Rather, it is to point to the **principle** of action that Bernie must implicitly accept if he does falsify the data. Once Bernie looks at his situation in terms of this broader principle, he will likely find it much more difficult to find falsifying the data acceptable than if he asks only what are the likely consequences of doing as Alex requests.

Further Reading

1. Bok, Sissela, Lying: Moral Choice in Public and Private Life (New York: Vintage Books, 1978).
2. Jaksa, James and Michael S. Pritchard, Communication Ethics: Methods of Analysis (Belmont, CA: Wadsworth, 1988). [See especially Ch. 7, "Groupthink" and Ch. 8, "The Challenger Disaster." Also, see Chs. 4-6 on methods of ethical analysis and justification.]
3. Janis, Irving, Groupthink, 2nd ed. (Boston: Houghton Mifflin, 1983), esp. pp. 14-47.

Ethics Case Study 2: A Tourist Problem

This and the other case studies presented in this book were developed through the National Science Foundation, with Grant No. DIR-8820837, headed by Michael S. Pritchard, Director, Center for the Study of Ethics in Society. These case studies are part of the public domain and are included here verbatim. Additional case studies and resources on engineering ethics can be found at http://ethics.tamu.edu/.

A Tourist Problem

I

Marvin Johnson is Environmental Engineer for Wolfog Manufacturing, one of several local plants whose water discharges flow into a lake in a flourishing tourist area. Included in Marvin's responsibilities is the monitoring of water and air discharges at his plant and the periodic preparation of reports to be submitted to the Department of Natural Resources.

Marvin has just prepared a report that indicates that the level of pollution in the plant's water discharges slightly exceeds the legal limitations. However, there is little reason to believe that this excessive amount poses any danger to people in the area; at worst, it will endanger a small number of fish. On the other hand, solving the problem will cost the plant more than $200,000.

Marvin's supervisor, Plant Manager Edgar Owens, says the excess should be regarded as a mere "technicality," and he asks Marvin to "adjust" the data so that the plant appears to be in compliance. He explains: "We can't afford the $200,000. It might even cost a few jobs. No doubt it would set us behind our competitors. Besides the bad publicity we'd get, it might scare off some of tourist industry, making it worse for everybody."

How do you think Marvin should respond to Edgar's request?

II

No doubt many people in the area besides Marvin Johnson and Edgar Owens have an important stake in how Marvin responds to Edgar's request. How many kinds of people who have a stake in this can you think of? [E.g., employees at Wolfog.]

III

Deborah Randle works for the Department of Natural Resources. One of her major responsibilities is to evaluate periodic water and air discharge reports from local industry to see if they are in compliance with antipollution requirements. Do you think Deborah would agree with the Plant Manager's idea that the excess should be regarded as a "mere technicality"?

IV

Consider the situation as local parents of children who swim in the lake. Would they agree that the excess is a "mere technicality"?

V

A basic ethical principle is "Whatever is right (or wrong) for one person is right (or wrong) for any relevantly similar persons in a relevantly similar situation." This is called the **principle of universalizability**. Suppose there are several plants in the area whose emissions are, like Wolfog Manufacturing's, slightly in excess of the legal limitations. According to the principle of universalizability, if it is right for Marvin Johnson to submit an inaccurate report, it is right for all the other environmental engineers to do likewise (and for their plant managers to ask them to do so). What if all the plants submitted reports like the one Edgar Owens wants Marvin Johnson to submit?

VI

Now that you have looked at the situation at Wolfog from different perspectives, has your view of what Marvin Johnson should do changed from your first answer?

[This case is an adaptation of "Cover-up Temptation," one of several short scenarios in Roger Ricklefs, "Executives Apply Stiffer Standards Than Public to Ethical Dilemmas," The Wall Street Journal, November 3, 1983.]

Commentaries

Kenneth L. Carper, School of Architecture, Washington State University

It is interesting to notice the language people use to justify unethical behavior. Plant Manager Edgar Owens refers to overlooking "mere technicalities," when he really means breaking established laws. He requests Marvin Johnson to "adjust" the report, when he really intends for Johnson to falsify scientific data.

The falsification of data is viewed by scientists and engineers to be an extremely serious breach of ethics. Marvin Johnson is being asked to compromise one of the most important moral concepts in science, truthfulness in reporting of scientific measurements. Should he consent to a false report, and should the incident come to light, his own personal career will be in grave jeopardy. The scientific and engineering community cannot survive unless its members can trust one another to present data truthfully.

Yet, Marvin Johnson finds himself in a very difficult position. His manager has raised the question of loyalty. The implication is that truthfulness will damage the company; fellow employees will suffer. Competitors will profit at the expense of Wolfog Manufacturing. The arguments given by Edgar Owens can be quite persuasive, and they are all too familiar in the corporate setting (Nelson and Peterson 1982). Regulations are often seen to be unrealistic or arbitrary. The assumption is often made that competitors must be falsifying data to meet these unrealistic expectations, so it is only wise business practice to do what everyone else is doing.

Much has been written about the pitfalls of misguided loyalty. While principled loyalty can be a commendable virtue, misguided loyalty has been responsible for many, many tragic moral disasters. When loyalty to a corporation, or a government, or an individual, requires the sacrifice of fundamental moral principles, such loyalty is not a virtue.

Engineers who find themselves in stressful situations like this should refer to their professional Code of Ethics. This can be a helpful, tangible tool in negotiations with their employers. (Carper 1991, Davis 1991). Certain fundamental ethical principles are embodied in the Codes of Ethics adopted by professional societies, and the embattled engineer can point to these principles, stating that his or her career as an engineer requires adherence to these principles. What Johnson is being asked to do is a violation of the canons of his profession.

The principle of universalizability is introduced in this case study. Immanuel Kant's "categorical imperative" provides this guidance:

Act only according to that maxim by which you can at the same time will that it should become a universal law.

In this case, Johnson should not write an "adjusted report" unless he is truly willing to accept similar actions by all his colleagues in the scientific and engineering community when confronted by similar situations and similar pressure from their employers. Should Johnson consent to Edgar Owens' request, later self-analysis of his actions will bring the crisis of conscience experienced by others who have compromised their values in the interest of misguided loyalty.

One relevant example is the B. F. Goodrich case involving data falsification on critical brake and wheel assembly testing for Air Force attack aircraft (Martin and Schinzinger 1989, p.58). The firsthand account provided by Kermit Vandiver, a B. F. Goodrich employee, is very enlightening (Vandiver 1972).

Deborah Randle, the engineer who works for the Department of Natural Resources, will most certainly evaluate reports from the various corporations with the principle of universalizability in mind. How else can someone charged with global responsibility operate, and remain impartial? False data will be absolutely unacceptable to Randle. Again, engineers simply must be able to trust each other.

Should an unethical report be discovered, not only will Johnson's reputation be irreparably damaged, but the impact on Wolfog Manufacturing will also be significant. The case of emissions test data falsification by the Ford Motor Company shows the damage such behavior can do to a corporation (Martin and Schinzinger 1989, pp. 163164). A review of the Ford case illustrates the fact that compromising ethics in the interest of loyalty can actually result in great damage to the very employer one is trying to protect.

It seems that Marvin Johnson has some thinking to do. It is probably not yet time to "blow the whistle" publicly. There are some moral principles and procedures involved in proper whistleblowing, and Johnson has not yet exhausted his avenues within the corporation (Elliston et al 1985). Indeed, Johnson has an excellent opportunity to provide some moral leadership to his colleagues by speaking out on the issue of scientific truthfulness. But engineers simply must refuse to work for corporations that place profit above scientific honesty. If Edgar Owens represents the moral stature of the Wolfog corporate management, then Wolfog Manufacturing is not a healthy environment for an honest engineer.

Suggested Readings:

1. Carper, Kenneth L. 1991. "Engineering Code of Ethics: Beneficial Restraint on Consequential Morality," Journal of Professional Issues in Engineering Education and Practice, American Society of Civil Engineers, New York, NY, Vol. 117, No. 3, July, pp. 250257.
2. Davis, Michael 1991. "Thinking Like an Engineer: The Place of a Code of Ethics in the Practice of a Profession," Philosophy and Public Affairs, Princeton University Press, Princeton, NJ, Vol. 20, No. 2, Spring, pp. 150167.
3. Elliston, Frederick, J. Keenan, P. Lockhart and J. van Schaick 1985. Whistleblowing Research: Methodological and Moral Issues, Praeger Publishers, New York, NY, pp.133161.
4. Martin, Mike W. and R. Schinzinger 1989. Ethics in Engineering (2nd edition), McGrawHill, Inc., New York, NY, pp. 58, 163164, 176.
5. Nelson, C. and S. R. Peterson 1982. "The Engineer as Moral Agent," Journal of Professional Issues in Engineering, American Society of Civil Engineers, New York, NY, Vol. 108, No. 1, January, pp. 15.
6. Vandiver, Kermit 1972. "Why Should My Conscience Bother Me?" from In the Name of Profit, by Robert L. Heilbroner, Doubleday and Company, Inc., Garden City, NY, pp. 331.

Joseph Ellin, Department of Philosophy, Western Michigan University

I

This case involves a violation of environmental regulations which may be more "technical" than real. Wolfog Co is faced with $200,000 unnecessary expenses to prevent small excess omissions which are not believed to be harmful to anyone but a few fish. The obvious course here is for Wolfog to apply to the DNR for a variance. Their lawyers can try to convince the DNR that the slight excess poses little danger. If they don't get the variance, they'll have to conform, or go to court; though all this will probably cost Wolfog more than the cost of compliance.

However there's nothing to be done on an individual basis. Manager Edgar Owens should not expect engineer Marvin to "adjust" the data and Marvin shouldn't do it. Edgar's reasoning is self-serving: if he's worried about image and tourism he should comply with the regulations. It may well be true that if Wolfog has to spend the $200,000 which they can't afford, they're in trouble, but the answer, if there is an answer, is not to fake data.

II

This might be one of those cases in which most people are better off if the law is violated rather than obeyed. Such situations are probably more common than realized. It's not the discharge itself which does any harm, but the fact that it's not in conformity to the regulations, since this creates the image problem and scares away the tourists. This obviously makes an excellent case for loosening the regulations: regulations should not be more onerous than necessary to achieve their purpose. The more people who have a stake in economic development, the more likely it is that this case will be heard by the authorities.

III

Whether Deborah, the DNR water quality official, would agree that the violation is a "mere technicality," depends on Deborah. We don't know enough about her; if she's a radical environmentalist, she thinks zero dead fish is the only tolerable condition, and no cost is too great to achieve it. She also may think there is no such thing as a technical violation: a violation is a violation, may be her enforcement motto. One might take the view that if she thinks this, she shouldn't be in her position, but perhaps her boss thinks so too. Perhaps this is the motto of the entire DNR, which if it is shows something about the irrationality we've gotten ourselves into.

IV

Would the parents agree that the violation is merely technical? Probably not; the local parents have been whipped up by the environmentalists and the media to think that any drop of anything is dangerous. They want jobs, economic progress, low taxes, low prices, and a pristine environment as well, (who doesn't?) and they are not willing or able to understand the issues involved. And they vote.

V

So given this hypothetical gloomy situation, is the over-all best solution that Marvin should just fake the data? One might make such an argument from a narrow act utilitarian point of view, but for all sorts of reasons including long-range utility it isn't right for anyone to submit a fake report, so the question whether everyone might do so is purely hypothetical. Another question would be, if it's right to grant a variance to Wolfog, is it right to grant a variance to every plant? And the answer would be yes, which is not an argument not to grant the variance to Wolfog, unless there is a comparable compelling reason at the other industries (for example, it might not cost everybody $200,000 to clean up). If there is, then the DNR is within its rights in denying the variance. If all the factories together produce a total discharge that is dangerous, the situation changes by that fact. But if there are no other plants in Wolfog's situation, then the so-called principle of universalizability should not be used as an excuse to impose hardships on one firm without any compensating gain for anyone except the few fish.

VI

Marvin shouldn't fake the data. The rest is up to the people at Wolfog.

Ethics Case Study 3: Property

This and the other case studies presented in this book were developed through the National Science Foundation, with Grant No. DIR-8820837, headed by Michael S. Pritchard, Director, Center for the Study of Ethics in Society. These case studies are part of the public domain and are included here verbatim. Additional case studies and resources on engineering ethics can be found at http://ethics.tamu.edu/.

Property

I

Derek Evans used to work for a small computer firm that specializes in developing software for management tasks. Derek was a primary contributor in designing an innovative software system for customer services. This software system is essentially the "lifeblood" of the firm. The small computer firm never asked Derek to sign an agreement that software designed during his employment there becomes the property of the company. However, his new employer did.

Derek is now working for a much larger computer firm. Derek's job is in the customer service area, and he spends most of his time on the telephone talking with customers having systems problems. This requires him to cross reference large amounts of information. It now occurs to him that by making a few minor alterations in the innovative software system he helped design at the small computer firm the task of cross referencing can be greatly simplified.

On Friday Derek decides he will come in early Monday morning to make the adaptation. However, on Saturday evening he attends a party with two of his old friends, you and Horace Jones. Since it has been some time since you have seen each other, you spend some time discussing what you have been doing recently. Derek mentions his plan to adapt the software system on Monday. Horace asks, "Isn't that unethical? That system is really the property of your previous employer." "But," Derek replies, "I'm just trying to make my work more efficient. I'm not selling the system to anyone or anything like that. It's just for my use -- and, after all, I did help design it. Besides, it's not exactly the same system – I've made a few changes." What follows is a discussion among the three of you. What is your contribution?

II

Derek installs the software Monday morning. Soon everyone is impressed with his efficiency. Others are asking about the "secret" of his success. Derek begins to realize that the software system might well have company-wide adaptability. This does not go unnoticed by his superiors. So, he is offered an opportunity to introduce the system in other parts of the company.

Now Derek recalls the conversation at the party, and he begins to wonder if Horace was right after all. He suggests that his previous employer be contacted and that the more extended use of the software system be negotiated with the small computer firm. This move is firmly resisted by his superiors, who insist that the software system is now the property of the larger firm. Derek balks at the idea of going ahead without talking with the smaller firm. If Derek doesn't want the new job, they reply, someone else can be invited to do it; in any case, the adaptation will be made.

What should Derek do now?

III

Suppose Horace Jones is friends with people who work at the smaller computer firm. Should he tell them about Derek's use of the software system?

Commentaries

Neil R. Luebke, Department of Philosophy, Oklahoma State University

The general problem area raised by this case is ownership and use of technical knowledge. One question might be phrased, "What is the right of the individual engineer to specific items of technical knowledge which he/she came to possess because of and while in the employ of a firm?" However, there are additional, more broadly ranging, questions such as, "How, through communicating or (mis)using technical knowledge gained in a previous employment, might an engineer cause a current employer considerable trouble and expense?" and "To what extent is an engineer responsible for the firm's use of his/her technical knowledge?" Particular cases dealing with intellectual property, as in the present instance, are often complicated by a number of legal considerations. In any event, Derek would be well-advised to seek the counsel of his firm's lawyers before proceeding and to urge his superiors to do the same.

There are at least two major gaps in the case description that must be filled before concrete, detailed advice could be given to Derek. One is the matter of ownership, and the particular type of ownership, of the innovative software system. The other is whether Derek's new firm has a license to use the original software system. Regarding the first, the case description does not explicitly state "but strongly suggests" that the software system belongs to Derek's former firm. It is said to be the commercial "Lifeblood" of the firm, Derek and other developed it while employed by the firm (presumably as part of their jobs), and

Horace's remark "That system is really the property of your previous employer" is not challenged but apparently accepted by Derek. The fact that Derek did not sign an explicit agreement with his former employer may be immaterial. The former employer might have a company policy governing ownership. [My university has a detailed policy governing the ownership of patents and copyrights developed by university employees. No employee "signs" the policy and it went into effect, valid from its adoption date, years after I first joined the university faculty.] Then, too, there are legal precedents regarding design work done by employees while under hire to do such work. Derek's assumption that his helping to design the system gives him some sort of "right" to use it or change it may be dangerously flawed. It is remotely possible that Derek had and retains some sort of ownership right in the system, but there is nothing positive in the case description to suggest so.

There is also the question of Derek's current firm's license to use the software system, for they obviously do not own the original version. The case description suggests that the current firm has not purchased a license to use the original system, but wants Derek to replicate the whole system, with a few modifications, for extensive company use and (claimed) ownership. Such a move would seem to be a violation of copyright by the firm. It is not clear that Derek himself has a license to use the system – he may be "pirating" it. There are situations in which a company buys a license to use a software system and, as part of the purchase, also buys access to and use of the "source code" for the system, legally permitting the company to modify the system for its own needs. If this were the case – but there is nothing of the sort stated or implied in the description – there would be no problem in introducing the modification and the firm is fortunate to have a person with Derek's experience. (There may still be a problem with "owning" the modified system.) More commonly, the license to use a system prohibits the licensee from modifying the system (which would at least void the warranty) or distributing it to other parties; the software supplier retains the right to service, updates, or otherwise control modifications in the system. The user's rights and their limits are usually spelled out in legal detail in the purchase agreement and warranty.

As both Derek and his new employers could be exposed to a major lawsuit, Derek should insist that his superiors obtain legal counsel before the situation develops further. Derek may already be guilty of using proprietary information.

[Perhaps a few background remarks are in order here concerning the topics of patents and copyrights, trade secrets, and agreements that might be signed regarding maintenance of confidentiality or other intellectual property rights. The United States Constitution, Article I, Section 8, Clause 8. There the Constitution provides that "Congress shall have the power… to promote the progress of science and useful arts, by securing for limited times to authors and inventors the exclusive right to their respective writings and discoveries." The United

States was one of the first countries to develop a patent and copyright system. Since the country was founded at the beginning of the modern industrial revolution, there was a concern on the part of the Constitution writers that creative individuals be protected and encouraged in their efforts to develop inventions or to produce artistically. The federal Patent Act grants a limited monopoly to a patent-holding inventor that gives him a 17-year right to exclude others from using, making, or selling his invention in the United States. Once a patent is applied for, it becomes a form of personal property; it may be bought, sold, traded, leased, and licensed for the use of others, and so on. Not every invention is patentable; it has to meet certain standards, usually standards of novelty, utility, inventiveness, and what is sometimes called subject matter, that is, there are certain things which cannot be patented, such as mathematical formulas or managerial techniques, whereas other things – such as actual devices, designs for processes, and chemical formulas – may be patented. A copyright is similar to a patent in that it grants an exclusive right to the holder to publish, sell, etc., plays, music, textbooks, photographs, and other such material. The extent of protection is usually for longer for a copyright than for a patent. To infringe a copyright or to manufacture illegally, say, a patented device would be grounds for a lawsuit by the owner of the patent or the copyright against those who violated the right.

There is another category of protected information usually called "confidential information," "proprietary information," or more commonly "trade secrets." For a variety of reasons, companies may prefer to use the trade secret route to protect information, designs, formulas, and possibly even computer programs developed within their firms. One reason is the expense associated with patenting. Another reason has to do with the fact that a patented or copyrighted item becomes a matter of public record and hence is publicly accessible. While at the time of this writing there is no national trade secret law, there are a number of statutes on the books of various states and a large number of court decisions protecting the rights of a firm to maintain its own trade secrets. Companies have successfully sued each other for trade secret theft. (Surely <u>they</u> were never parties in signing any ownership agreement!) One process by which trade secret theft sometimes occurs is the hiring of an employee from another firm and then putting that employee in a position to use specialized knowledge obtained from the previous firm. A trade secret can be any specific piece of information – a list of customers, a mathematical formula, a chemical formula, a process design, and so on – that gives the company holding it a competitive advantage in the marketplace and has been identified or treated by the company as confidential. Generalized skills that an employee may pick up on the job, such as a skill in using certain programming techniques or a skill in the conducting some type of chemical analysis, would not be regarded as trade secrets but rather as general knowledge that becomes part of the employee's overall ability.

In order to help the company in protecting trade secrets (an activity that is extremely important in more technological firms) as well as in alerting employees to policies governing patents and copyrights, employees are often requested to read and sign an employment agreement. These agreements explain the company's policy regarding patents, copyrights, and trade secrets and usually call attention to the employee's obligation to continue to maintain confidentiality indefinitely, not just during the time of employments with the company.]

Let us suppose that the company, at Derek's insistence, does carry out negotiations with the smaller company producing the original software. Derek's ethical commitments to his previous and his current employers would not be violated. He will have kept faith with his previous employer, and he cannot be charged with knowingly doing anything to damage his current employer. His current firm may be able to work out some kind of relationship with the previous employer and possibly even come to a business arrangement for the marketing of the adaptation for even more commercial profit.

On the other hand, let us suppose that the company does not talk with the supplier of the software and insists on going ahead with or without Derek in making the adaptations company-wide. Let us also suppose, as is likely to be the case, that at least someone in Derek's previous company finds out about the new use of the software system. Derek's new employer, as well as possibly Derek himself, could be confronted by a lawsuit having to do with the infringement of copyright. Even if the lawsuit is not successful, the legal action could prove costly for Derek's employer. It certainly might besmirch his reputation, both within the company and outside of it. Derek should at least do what he can to urge his superiors to seek legal advice in this matter. His professional responsibilities to his current employer as well as to his previous employer would call for no less. If his superiors are too pig-headed to seek and listen to legal counsel on this matter, Derek would probably be better off working for a more enlightened firm¾at least one that would not ask him to act illegally.

John B. Dilworth, Department of Philosophy, Western Michigan University

This is one of the few cases where the specific legal provisions governing the matters at issue are of primary importance in clarifying and resolving the problems. In the case of software, some basic points about copyright law, and some related matters concerning software licensing, are vital to understanding the case, and to distinguishing it from other cases of ownership or rights as they apply to employees. Therefore these legal provisions will be spelled out as an integral part of this commentary.

1. Copyright in software is treated under current U.S. law as being essentially similar to copyright in any literary or creative work. In all these cases, one acquires initial ownership of the copyright simply by being the author of the work in question. (Several persons may

jointly author a work and so jointly hold copyright to it.) The government (through the Copyright Office, Library of Congress) does provide a Registration of Copyright mechanism. This does not create ownership, but instead officially acknowledges that it already exists. Typically one submits a manuscript, whether of a novel, movie script or source code for a program, as evidence of one's authorship/copyright.

Authorship as discussed above is subject to the following important qualification. The author of a work might prepare it as a "Work Made for Hire" (defined as a work prepared by an employee within the scope of his/her employment), and explicitly state this on the Copyright Registration form. In this case, the copyright statute provides that the employer rather than the employee is considered as the author (and hence as the copyright holder). However, it is important to note that, in the absence of any such explicit acknowledgement by an author that the work was "Made for Hire," the normal assumption would be that the actual author/s hold copyright to the work, unless other substantive evidence could be produced to prove that it was "Work Made for Hire."

In the current case, we are explicitly told that Derek was never asked by the small computer firm to sign an agreement that software designed during his employment there becomes the property of the company. If he signed no such agreement, nor (as we may consequently assume) specified that his work was a "Work Made for Hire" in any copyright registration application, then legally he would have a strong presumptive case that he was (and still is) the copyright owner of the software in question. (The possible complication that he was the primary, but not the only, contributor to the software will be considered later.)

It might be objected that in most cases of employer-employee relations, if one works for someone then they own the products of one's labor. This is broadly true, but creative works falling under the copyright laws work differently. A familiar example in higher education is the fact that professors retain copyright in their books or papers even if they were hired to carry out such creative research.

In the present case, the fact that Derek was indeed working for a computer firm while preparing the program etc. is not sufficient to establish copyright ownership by the firm. For in the case of software copyrights, there are other rights or permissions to use the software which the firm will have acquired as a result of Derek's activities, which are fully adequate for their business purposes and which justify their hiring and compensating of Derek for his work. They get broadly what they want, but it is rights and permissions to use the software which they get, rather than ownership of it. (Recall that they could have had ownership too, but neglected or elected not to take the necessary legal steps to secure it.) Here is a brief discussion of rights and licensing in relation to copyright, to help clarify these matters.

2. Everyone is familiar to some degree with literary and movie rights, for example that a producer may have to pay a novelist a large sum to get the movie rights to a book. These rights give the movie producer the right to produce a film version of the novel, but do not in any way transfer the copyright (or ownership) of the novel to the producer. Similar considerations apply to software too: acquiring the right to make certain uses of software does not transfer its ownership. Derek's firm acquired rights to use his software system for customer services in virtue of his being employed by them to produce the software, but the firm does not thereby acquire property rights in the software.

More distant still from ownership considerations are issues about licensing. Almost all actual software contracts involve some kind of software licensing, in which the copyright owner gives permission to one or more licensees to make certain kinds of use of the software. Though an exclusive license is possible, most licenses are of a non-exclusive kind, so that many different licensees could make similar uses of the software without violation of their contracts. Clearly in such cases there is no question of any transference of ownership in the legal arrangements.

3. In the present case we are given no specific information about what rights or licensing arrangements were in force between Derek and his original small computer firm, but these can be reasonably inferred from the conduct of the parties. Minimally his firm needed from Derek a perpetual, non-exclusive license to use and modify his source code for the software. Then they could use the software indefinitely, and modify it at any time in the future as changes became desirable. However, unless there was a written contract in existence (signed by the firm and Derek) in which Derek granted the firm an exclusive license to use the software, Derek is free at any time to license the same software (whether or not he chooses to make changes in it) to anyone else, and under any terms he wishes.

The implications for the present case are clear. Derek as the copyright holder can use or modify his software for use in his new larger computer firm in any way he pleases, with or without discussing it with his former employer. What is more, his new employer cannot claim ownership of the software, because it was developed prior to Derek's current employment rather than as part of his current design work.

However, Derek would certainly be wise to come to some explicit agreement with his new firm about how he would allow them to use the software. In effect, the new firm wants Derek to produce a customized version of the software for them, and he could agree to do this as part of his regular compensation, while also negotiating a monthly or yearly licensing fee in return for granting them appropriate rights to use the software. Or to simplify things, Derek might be tempted to sell them the package outright for a suitable compensation, in which case there would be an actual transference of ownership of the package.

4. We are told that Derek was a "primary contributor" in the original development of the software. This suggests the possibility that he may jointly hold the copyright to the software with one or more other designers. How would this affect the case? Generally, joint ownership allows each owner to exercise all rights of ownership, except for those whose exercise would materially affect the rights of the remaining owners. (Commonplace examples include such matters as joint ownership of a home or bank account.) In the present case, this means that Derek is free to grant non-exclusive licenses to use the software to anyone (but not to everyone), since other joint owners would not be thereby prevented from exercising similar rights. On the other hand, Derek should refrain from attempting to grant an exclusive license, or from attempting to sell the software outright, because both of these actions would materially affect the interests of any other joint owners.

Does Derek have any moral obligation to contact his former employer, or his co-workers there, before exercising his legal rights as detailed above? First, it is prudent for anyone in business to stay on good terms with both present and former associates. In the interests both of common courtesy and of safeguarding his own career, Derek would be well advised to explain his actions and his view of the case to anyone who might otherwise resent or be annoyed by them, including his former associates and friends.

Second, if there are indeed co-authors from his previous firm whom Derek could contact, he should do so. A basic principle of legal ethics, assumed in contract law, is that parties who enter into a contract or agreement are thereby obligated to make a good-faith effort to carry out the terms of the contract, both explicit and implicit. Co-authorship, as with other forms of joint ownership, could appropriately be viewed as requiring that one should keep co-authors informed of one's actions with respect to the joint property, even if this is not explicitly spelled out in a written agreement between the co-authors.

Joseph Ellin, Department of Philosophy, Western Michigan University

I

This is a debate over who owns a software system, the company or the designer. Horace says the system is "really" the property of the company, but Derek, the designer, claims to possess certain property rights in it. My contribution to the ensuing discussion would be to say: "Gentlemen. Questions of property are determined by law, not philosophy. It is true that some philosophers, such as Locke and Nozick, think there's such a thing as property apart from law; but this opinion is untenable, as no rational criteria can be provided by which 'natural' property can be determined. (For instance, Locke says that I own anything with which I 'mix my labor'; but what is that? If I build a fence around a forest, does that mean I own the entire forest? Or only the land under the fence? Or the fence itself and nothing else? And if I own the fence, do you have a right to climb over the fence to get into the forest?) Whether the system is or is not the property of the previous employer depends on what the law says. It's not a moral question whether Derek or his prior employer have legal control over the work Derek did there. This may or may not depend on any agreement Derek signed, or didn't sign. So Derek should consult his, or the company's, lawyer and determine what his rights are.

Derek's arguments are feeble rationalizations for his desire to fiddle with the software he invented. He'd be more honest to say, "Look, if I had signed my rights away, my hands would be tied. But I never did, and the previous company didn't seem to care. So now I'm free as a bird to do what I want with this system. That's my understanding of my legal rights, and I intend to exercise them fully. If I'm wrong, they can sue me.'"

II

As the case develops, things go farther than intended and now Derek's new employer not only wants him to make greater use of the software system than he initially thought would be right, but claim to own it themselves. Derek has gotten himself into a moral pickle and he wants someone to rescue him from it. He signed an agreement with his new company that his work belongs to them. He then revised the work he did for the first company, half-thinking that he really shouldn't; and now he regrets that the new company claims that the whole thing is theirs! It's really too late for him to conclude that the old company is being treated badly, since he's the one who made it happen. The only remedy here is for the first company to sue the new employer and have the court determine the legal property rights.

III

Should Horace, Derek's friend, tell the old company that Derek is using the software? Why not? There are no secrets involved in this, unless Horace is under a pledge of confidentiality, which wasn't stated. The smaller company needs to know that their systems are at risk unless they secure legal title to them; it's surprising that they have never been told this before by their lawyers! (Maybe they need new lawyers).

Michael S. Pritchard, Department of Philosophy, Western Michigan University

It might seem that this case is basically about law rather than ethics. Clearly it does raise a number of legal questions. However, there is a strong ethical dimension as well. Derek's desire to adapt the software program to his new job circumstances seems innocent enough. But the fact that his new employer required him to sign a software agreement that what he designs becomes company property should have alerted him to a potential problem. Although Derek did not sign a similar agreement with his previous employer, this does not conclusively settle the question of ownership. Others were involved in the initial design.

At the very least, Derek should have inquired about the ownership matter prior to adapting the software to his purposes. This would not only protect his current employer from a potential law suit (should the previous employer choose to sue), it would also evidence respect for the interests of his previous associates. Carelessly placing one's employer at legal risk is both an ethical and a legal concern. Indifference to the interests of his previous associates is an ethical concern, unless we can assume that Derek is estranged from them (and even if he is, there might have been an implicit understanding about the disposition of the software). After all, Derek is very possibly legally entangling the "lifeblood" of his previous employer, given his current employer's apparent desire to claim ownership of its employees' software designs.

It might be objected that Derek did not know that his new employer would use all means at its disposal to adapt the software system throughout the company. True, but his having to sign an ownership agreement should have put him on alert.

It seems clear from the case that Derek bore no special animosity against his previous employer and associates. Now, to his regret, he has become involved a legal and ethical quagmire. Perhaps a careful investigation of law can clarify the legal rights involved in this case, but the ethical concerns cannot be handled so readily. So, I conclude that Derek should have proceeded with greater caution, heeding the concerns of Horace. A call to his previous employer before adapting the system might have avoided these problems.

Essential Mathematics and Physics Skills

Significant Digits and Rounding

Significant Digits

One way of thinking about significant digits is to say, "I am reasonably confident that these numbers are correct and relevant."

You often use significant digits whether or not you are really thinking about it. For example, say that you had to divide 3.15 by 6.02. To be completely accurate, we could say:

$$\frac{3.15}{6.02} = 0.523255813953488 \ldots$$

This, however, is pretty ridiculous. Even though it's "correct," it's also irritating. Most likely nobody cares whether the answer is 0.523255813953488 or 0.523255813953489 or 0.523255813953487. That isn't to say that there aren't some instances where precision is important, but generally our original measurements are only so accurate to begin with, so reporting such precision in a calculation is irrelevant and unnecessary.[*] (If you are interested in this kind of thing, you can look up "Geometric Tolerance and Design," or GD&T.)

In any case, the "proper" way to express something like this would be to say

$$\frac{3.15}{6.02} \approx 0.523$$

Before we get into a discussion of how many significant digits to include in calculations, however, it is important to define what we mean by significant digits. Here are some general rules and examples:

---

[*] For a more detailed discussion of accuracy and precision, see the chapter on Lab Measurements and Error Analysis.

Rule	Example	Number of Significant Digits
• Any non-zero digit is significant.	549	3
	1.9423	5
• Any number between two non-zero digits is significant.	30229	5
	601	3
	5.003	4
• Zeros that appear before the first non-zero digit are not significant	0.0000032	2
	0.05	1
	0.00000507	3
• For values represented with decimal points, any zeros after a non-zero value are significant.	0.50	2
	0.500	3
	5.50	3
	5.5000	5
	500.0	4
• For values not represented with decimal points, the "trailing zeros" are ambiguous.	500	1-3
	20930	4-5
	1800	2-4
	6300	2-4

Finally, remember that *an exact number has an infinite number of significant digits*. Consider that the perimeter P of a square of width w is given by

$$P = 4w$$

The number 4 in this equation is an exact number. There are four sides of the square, so the perimeter is the width of one side times four. There is no error in the value 4, so it is considered to have an infinite number of significant digits.

Rounding

A discussion of significant digits is invariably followed by a discussion of rounding. In engineering, we are a little more careful with rounding than you may be used to. We will use the following conventions when discussing rounding (assume that we are rounding to 3 significant digits):

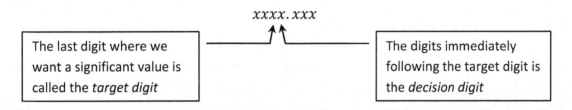

The last digit where we want a significant value is called the *target digit*

The digits immediately following the target digit is the *decision digit*

Rule	Example	Rounded to 3 Significant Digits
• If the decision digit is greater than 5, round the target digit to the next highest number	2346.2	2350
	92.76	92.8
	2.3863	2.39
	389.98	390.[†]
• If the decision digit is less than 5, leave the target digit as it is	905.3	905
	0.02342	0.0234
	6342.2	6340
	370.2	370.[†]
• If the decision digit is equal to 5 and is followed by additional non-zero digits, round the target digit to the next highest number	2975.342	2980
	7.63523	7.64
	98.8500001	98.9
	100.59	101
• If the decision digit is equal to 5 and is followed by zero digits or not followed by anything... ○ if the target digit is even, leave it as is ○ if the target digit is odd, round it up to the next even number	293.50000	294
	293.5	294
	294.50000	294
	294.5	294
	0.009895	0.00990
	0.009885	0.00988

[†] Notice that here we must use a decimal point to indicate that all three values, including the 0, are significant.

You should probably be familiar with all but the last rule, which may be new to you. The reason we have this seemingly unnecessarily complicated method is because rounding this way prevents statistical bias. If you always rounded up on a perfect 5, then your measurements would eventually tend a little higher than they really should be.

Significant Digits in Calculations

The number of significant digits in the result of a calculation is determined by the number of significant digits in the original data used in the calculation. Always use the lowest number of significant digits from the original data to determine the number of significant digits in a calculated result.

For example,

$$3.02 \cdot 9.002342 = 27.18707284$$

However, we can only be as precise as our most imprecise measurement. The two values we multiplied together have 3 and 7 significant digits, respectively. Therefore, our calculated value can have no more than 3 significant digits.

$$3.02 \cdot 9.002342 \approx 27.2$$

Remember that when you are multiplying by exact values that these exact values have an infinite number of significant digits. For example, if we measure one edge of a square and get 9.03 m, we can calculate the perimeter P as follows:

$$P = 4w = 4 \cdot 9.03 \text{ m} = 36.12 \text{ m} \approx 36.1 \text{ m}$$

Even though you may think that the number 4 only has one significant digit, remember that it in fact has an infinite number of significant digits. The value with the lesser number of significant digits is 9.03, which has three. Therefore, we round our calculation off to three significant digits.

That said, remember never to round off your calculations until you are done. For example, consider that we are adding together the perimeters of three different squares to calculate how much fencing we should buy (figures below are not drawn to scale).

$w_1 = 9.03$ m

$P_1 = 36.12$ m

$w_2 = 5.29$ m

$P_2 = 21.16$ m

$w_3 = 7.69$ m

$P_3 = 30.76$ m

If we rounded off P_1, P_2, and P_3 before adding them together, we would get

$$P_1 + P_2 + P_3 \approx 36.1 \text{ m} + 21.2 \text{ m} + 30.8 \text{ m} = 88.1 \text{ m}$$

If we do not round them off before adding them together (only rounding when we get a final solution), we would get

$$P_1 + P_2 + P_3 = 36.12 \text{ m} + 21.16 \text{ m} + 30.76 \text{ m} = 88.04 \text{ m} \approx 88.0 \text{ m}$$

You may not think that the difference in the two answers is significant, but it does not take much imagination to see how continually rounding off over and over again can drastically impact an eventual solution.

The best way to proceed with calculations is to never "clear" your calculator until you are at the final answer. Never write down intermediate steps as you go unless it's absolutely necessary. If you must do this, be sure and record lots of "extra digits" so that you don't lose them along the way. You can always round off later.

Binary and Hexadecimal Numbers

The Common Number System

Ask any child how old he or she is, and you are likely to get a response of, "This many!" with a proud display of the appropriate number of fingers. It is not a hard stretch to imagine that the first accounting was done with thumb, index, and the rest of the clan. And, mostly likely because we do have 10 fingers, we also have a base-10 number system.

In a base-10 number system, or a *decimal* system, we have 10 digits that we use to represent values: 0, 1, 2, 3, 4, 5, 6, 7, 8, and 9. If we need to express a higher number than 9, we have to use more than one digit. Our two-digit numbers range from 10 to 99. Our three-digit numbers range from 100 to 999.

In general, the highest number that you can express in a base-x number system with one digit is $x - 1$. Similarly, the highest number that you can express in a base-x number system with n digits is $x^n - 1$.

Number of Digits	Highest Number that can be Represented
1	$10 - 1 = 9$
2	$10^2 - 1 = 99$
3	$10^3 - 1 = 999$
4	$10^4 - 1 = 9999$

The Egyptians, Greeks, and early Chinese had a base-10 number system as we do today. Not all civilizations worked that way, however. The Mayans used a base-20 number system (perhaps because their warm climate didn't require shoes so they had the advantage of using both fingers and toes to count?). The Babylonians had a base-60, or sexigesimal, number system, which is why we have 60 minutes in an hour, 60 seconds in a minute, and 180 degrees in a triangle (3 x 60).

You may not have put too much thought into your number system since grade school, but to grasp what binary and hexadecimal numbers are, you must learn to think more formally about the decimal numbers you use every day. For example, recognize that the number 526 really means that we have five 100s, two 10s, and 6 ones, or units.

Binary and Hexadecimal Numbers

Decimal Number	"Units" Representation		Power Expansion
526	5 in the 100s place 2 in the 10s place 6 in the 1s (units) place	500 + 20 + 6	$5 \cdot 10^2 + 2 \cdot 10^1 + 6 \cdot 10^0$
1025	1 in the 1000s place 0 in the 100s place 2 in the 10s place 5 in the 1s place	1000+000+20+5	$1 \cdot 10^3 + 0 \cdot 10^2 + 2 \cdot 10^1 + 5 \cdot 10^0$
73.4	7 in the 10s place 3 in the 1s place 4 in the 10ths place	70 + 3 + 0.4	$7 \cdot 10^1 + 3 \cdot 10^0 + 4 \cdot 10^{-1}$
3.14	3 in the 1s place 1 in the 10ths place 4 in the 100ths place	3 + 0.1 + 0.04	$3 \cdot 10^0 + 1 \cdot 10^{-1} + 4 \cdot 10^{-2}$

Notice that when you're counting to determine which number goes in the exponent of the power expansion, you start counting at 0. For example, the "6" in "526" is in the **0**th position (and hence represented as $6 \cdot 10^0$), the "2" is in the **1**st place (represented as $2 \cdot 10^1$), and the "5" is in the **2**nd place (represented as $5 \cdot 10^2$). For negative exponents, you begin counting immediately at -1, -2, -3, and so on.

The Binary Number System

Computers do not have 10 fingers. They only know two things: True or False. Another way to say this is that they only know two numbers, 1 (true) or 0 (false). A number system that only has two numbers, 0 and 1, is called base-2, or *binary*.

Number of Digits	Highest Number that can be Represented in Binary	Decimal Equivalent
1	1	$2 - 1 = 1$
2	11	$2^2 - 1 = 3$
3	111	$2^3 - 1 = 7$
4	1111	$2^4 - 1 = 15$

The way numbers are represented in binary is similar to how they are represented in decimal:

Representation in Binary	Power Expansion	Decimal Equivalent
1	$1 \cdot 2^0$	1
10	$1 \cdot 2^1 + 0 \cdot 2^0$	2
11	$1 \cdot 2^1 + 1 \cdot 2^0$	3
100	$1 \cdot 2^2 + 0 \cdot 2^1 + 0 \cdot 2^0$	4
101	$1 \cdot 2^2 + 0 \cdot 2^1 + 1 \cdot 2^0$	5
11.1	$1 \cdot 2^1 + 1 \cdot 2^0 + 1 \cdot 2^{-1}$	3.5
1.01	$1 \cdot 2^0 + 1 \cdot 2^{-2}$	1.25
100010	$1 \cdot 2^5 + 1 \cdot 2^1$	34

Again, notice that for positive exponents, when you're counting to determine which number goes in the exponent of the power expansion, you start counting at 0. In the last example in the table above the sixth digit is actually associated with a power of 5 ($1 \cdot 2^5$), not a power of 6. For negative exponents, you again begin counting at -1, then -2, then -3.

Practice converting the following values from binary to decimal:*

a. 111
b. 10011
c. 1101
d. 1.011

Converting decimal numbers into binary requires thinking backwards. You need to determine how many of each power of two is in the decimal number. We'll first focus on whole numbers. It's good if you can remember at least the first seven powers of two, which are 1, 2, 4, 8, 16, 32, and 64.

Consider with the number 14. The biggest power of two that is in 14 would be 8. (The power 16 is too big, so we can't use it.) We determine that there is one 8 in the number 14, with 6 left over.

Binary and Hexadecimal Numbers

$$14 = 1 \cdot 2^3 + \ldots \text{ (6 left over)}$$

Now the biggest power of two that is in 6 would be 4. There is one 4 in 6, with two left over.

$$14 = 1 \cdot 2^3 + 1 \cdot 2^2 + \ldots \text{ (2 left over)}$$

The biggest power of two that is in 2 would be 2. There is one 2 in 2, with nothing left over.

$$14 = 1 \cdot 2^3 + 1 \cdot 2^2 + 1 \cdot 2^1$$

Fill in any "unused" powers:

$$14 = 1 \cdot 2^3 + 1 \cdot 2^2 + 1 \cdot 2^1 + 0 \cdot 2^0$$

And now express the decimal as a binary number:

$$14 \text{ (decimal)} = 1110 \text{ (binary)}$$

You can also try using a "slot" approach where you create a table similar to that below and put a check mark in the columns so that if you add all the values associated with the checkmarks together, you'll get your decimal number. Anything with a checkmark is a 1, and anything blank is a 0.

	2^6	2^5	2^4	2^3	2^2	2^1	2^0	**Binary**
	64	**32**	**16**	**8**	**4**	**2**	**1**	**Representation**
14				✓	✓	✓		1110
8				✓				1000
33		✓					✓	100001
19			✓			✓	✓	10011
68	✓				✓			1000100

Practice converting the following values from decimal to binary:[†]

    a. 9
    b. 12
    c. 22
    d. 76

One consequence of using binary numbers that you might not expect is the fact that you cannot express fractional values perfectly unless they can be written equivalently to a negative power of 2. For example 2^{-1} is ½ or 0.5, so 0.5 (decimal) can be written as 0.1 (binary). But to express 1/3, which is not a negative power of 2, you must use a repeating decimal: 1/3 = 1/4+1/16+1/64 + ... = 0.010101.... Just as in decimal, it is not possible to write 1/3 in binary without an infinite number of digits.

The Hexadecimal (Hex) Number System

To make things "easier," we often represent binary values for items like byte values, letters, or colors in a base-16 number system called hexadecimal, or hex. The "numbers" in hexadecimal are 0, 1, 2, 3, 4, 5, 6, 7, 8, 9, A, B, C, D, E, and F.

Hex "Number"	0	1	2	3	4	5	6	7
Decimal Value	0	1	2	3	4	5	6	7

Hex "Number"	8	9	A	B	C	D	E	F
Decimal Value	8	9	10	11	12	13	14	15

If you have a wireless router and have ever experimented with your WEP key, you have probably noticed that you're only allowed to use the characters 0 – 9 and A – F. This is not an arbitrary decision – it's related to hexadecimal.

Number of Digits	Highest Number that can be Represented in Hex	Decimal Equivalent
1	F	16 – 1 = 15
2	FF	16^2 – 1 = 255
3	FFF	16^3 – 1 = 4095
4	FFFF	16^4 – 1 = 65535

The way numbers are represented in hex is again similar to how they are represented in decimal:

Representation in Hex	Power Expansion	Decimal Equivalent
A	$10 \cdot 16^0$	10
B3	$11 \cdot 16^1 + 3 \cdot 16^0$	14
4F06	$4 \cdot 16^3 + 15 \cdot 16^2 + 0 \cdot 16^1 + 6 \cdot 16^0$	20230

The good news is we don't usually convert between decimal and hex. Usually, we're converting between hex and binary, and there's a really good technique to do this. (Most happily, we almost never represent fractional values in hex.)

For example, try converting 10111010111001 from binary to hex. First, starting at the right end of the number, group all the numbers by 4s (add zeros to the front of the number as placeholders as necessary).

$$0010 \quad 1110 \quad 1011 \quad 1001$$

Next, write the decimal equivalent of each grouping. The maximum value you can get for each grouping is $2^4 - 1 = 15$, which just so happens to be the biggest single digit number in hex.

$$2 \quad 14 \quad 11 \quad 9$$

Finally, write the equivalent hex term:

$$2 \quad E \quad B \quad 9$$

Therefore, 10111010111001 in binary is equivalent to 2EB9 in hex.

Representation in Binary	Group Expansion	Group Decimal	Hex Representation
1001011	0100 1011	4 11	4B
1110	1110	14	E
11100011	1110 0011	14 3	E3
10001	0001 0001	1 1	11

Practice converting the following values from binary to hex:[‡]

a. 1110011
b. 1101
c. 11011001
d. 1110010000011111

Other Notes

You will sometimes see numbers represented with a subscript, which indicates the base, such as 110010101_2 (binary), $\mathrm{F3FFEA}_{16}$ (hex), or 359_{10} (decimal).

Alternatively, numbers can be represented with a tag like BIN, HEX, or DEC to indicate a binary, hexadecimal, or decimal number, such as 110010101 BIN, F3FFEA HEX, or 359 DEC.

You can use the logic we've presented here to convert any non-fractional decimal number into any other base you like. Another common base that we haven't covered here is octal (base-8, or OCT), but it's the same idea.

Programmer Humor

See if these induce a rofl:

Q: There are only 10 kinds of people in the world: those who understand binary, and those who don't.[§]

Q: If only dead people understand hexadecimal, how many people understand hexadecimal?[**]

Q: Why do programmers often confuse Christmas and Halloween?[††]

---

[*] a.) 7 b.) 19 c.) 13 d.) 1.375
[†] a.) 1001 b.) 1100 c.) 10110 d.) 1001100
[‡] a.) 73 b.) D c.) D9 d.) E41E
[§] Since 10 in binary = 2, that means there are 2 kinds of people in the world.
[**] Since DEAD = $13 \cdot 16^3 + 14 \cdot 16^2 + 10 \cdot 16^1 + 13 \cdot 16^0 = 57005$, that means 57,005 people understand hex.
[††] Because 31 OCT = 25 DEC — that is, $3 \cdot 8^1 + 1 \cdot 8^0 = 24 + 1 = 25$

Scientific Notation and Unit Prefixes

Scientific Notation

Often in engineering and other scientific fields we have to deal with numbers that are very large or very small. Writing these numbers down in ordinary decimal form can be tedious and time consuming. As such, engineers and scientists commonly write numbers in "powers of ten" or "scientific notation."

For example, the mass of the Earth is 5,980,000,000,000,000,000,000,000 kg, but we don't want to have to write that every time we want to talk about the Earth's mass. Instead we write the number as a decimal number (5.98) multiplied by an appropriate factor of ten (10^{24}). Thus in scientific notation, the Earth's mass is 5.98×10^{24} kg.

On the other end of the spectrum, the mass of an electron is 0.000000000000000000000000000000911 kg. Again this is a tedious way to write this number. Once more we write this number as a decimal (9.11) multiplied by the correct power of ten (10^{-31}). Therefore in scientific notation, the mass of an electron is 9.11×10^{-31} kg.

If you're curious about how to get the number in the exponent of the 10, just remember that it's all about the decimal places. To make 5.98 equal 5,980,000,000,000,000,000,000,000, we have to move the decimal place to the *right* 24 times. That is, we have to multiply 5.98 by a million million million, or 10^{24}. Similarly, to make 9.11 equal 0.000000000000000000000000000000911, we have to move the decimal to the *left* 31 times, or multiply 9.11 by 10^{-31}.

Remember that for very large numbers, the exponent on the 10 is positive, and for very small numbers the exponent on the 10 is negative.

Note there is a specific format for numbers in "normalized" scientific notation: one digit, followed by a decimal point, followed by the rest of the digits, then multiplied by ten raised to a power. That is, we wouldn't write the mass of an electron as 0.911×10^{-30} kg or the mass of the Earth as 598×10^{22} kg, although these numbers are technically correct. To be in true scientific notation, there should be exactly one number in front of the decimal point.

To convert numbers back from scientific notation into "regular numbers," just move the decimal point the appropriate number of digits to the right or left, adding zeros as necessary. Again, remember that a positive exponent should make a large number and a negative exponent should make a small number. For example, 6.31×10^{4} will become 63100, while 6.31×10^{-4} will become 0.000631.

Don't let a negative sign in front of the number itself confuse you. That is, -6.31×10^4 just means that the number itself is negative, -63100, just as -6.31×10^{-4} means -0.000631.

Finally, remember special values of exponents, like 10^0, which just equals one. If we really want to write a number like 9.32 in scientific notation, we can say 9.32×10^0.

Scientific notation also solves the issue of significant figure ambiguity in numeric data. The way scientific notation is written you ALWAYS know how many significant figures there are in a number. For example, if you write 600, it is not clear whether you mean exactly 600, or something else entirely. By using scientific notation, you can tell your reader exactly what you mean. See an example in the table below.

Ambiguous Way to Write a Number	What You Really Mean	Intended Number of Significant Digits	Clearly Indicated in Scientific Notation
600	Exactly 600	3	6.00×10^2
600	Somewhere between 595 and 605	2	6.0×10^2
600	Somewhere between 550 and 650	1	6×10^2

Practice converting the following numbers into normalized scientific notation or into decimal notation:*

a. 589,700,000
b. 0.000789
c. -330,002
d. 4.29×10^5
e. 8.23×10^{-6}
f. -9.332×10^{-9}

Unit Prefixes

Another way to deal with very large or very small numbers is to use unit prefixes. In the metric system, unit prefixes can be used to substitute a single letter for specific powers of ten. These prefixes can be placed in front any unit in the metric system to express a larger or smaller unit (depending on the prefix used). There are units in the metric system for every conceivable type of measurement being made. You are probably familiar with the meter, gram, and second; but there are many more base units some of which are listed in the table to the right.

Unit	Symbol	Quantity
second	s	time
meter	m	length
gram	g	mass
Newton	N	force
Joule	J	energy
Watt	W	power
Coulomb	C	charge

Prefix	Symbol	Exponential Form
tera	T	10^{12}
giga	G	10^{9}
Mega	M	10^{6}
kilo	k	10^{3}
centi	c	10^{-2}
milli	m	10^{-3}
micro	μ	10^{-6}
nano	n	10^{-9}
pico	p	10^{-12}

There are many other units in the metric system, but these are some of the more common ones. Any of the above units can be used in conjunction with unit prefixes to simplify the scientific notation needed to express a value. Each unit prefix stands for a specific power of ten. The more common unit prefixes are given in the table to the left.

Scientific Notation and Unit Prefixes

Let us look at a couple of examples. Given a distance of 2200 meters, we could express that value as 2.2×10^3 m or, using the unit prefix "k" for 10^3 and "m" for meters, we can express the value as 2.2 km. Much simpler!

Don't be afraid to move the decimal place around. This is often necessary since unit prefixes are usually separated by orders of 10^3. Given a measurement of 5.6×10^{-5} Watts, we can change that to 56×10^{-6} Watts, and then express that as 56 µW. Again any metric unit can be combined with any unit prefix to express data in succinct detail.

Practice using unit prefixes with the following data:[†]

    a. 16,000 Watts
    b. 0.00095 meters
    c. 1.3×10^3 grams
    d. 6.7×10^{-7} Coulombs

---

[*] a.) 5.897×10^8 b.) 7.89×10^{-4} c.) -3.30002×10^5 d.) 429,000 e.) 0.00000823 f.) -0.000000009332
[†] a.) 16 kW b.) 950 µm c.) 1.3 kg d.) 670 nC

Unit Conversions

At one point the city of Snowmass Village had a sign welcoming visitors. The sign read:

Snowmass Village	
Established	1967
Elevation	8368
Population	1822
Total	12,157

The town leaders clearly had a sense of humor. We can't add years to feet to people and end up with some kind of total! Such a thing would be ridiculous!

Of course, what sometimes seems so obvious can somehow seem very strange and confusing when you're sitting in a cold engineering classroom, looking at a test you could've sworn you studied for, though for some reason all you can remember is the color of your Wii remote.

The purpose of this section, therefore, is to help you remember the basics about units, and how to get from one kind of unit to another.

Unit Systems

There are two different unit systems used in the United States: customary and metric. Customary and metric systems are different in that the metric system specifically relates its units by factors of 10 (a decimalized system), whereas customary systems are based on units that are often related arbitrarily. "Customary" and "metric" are actually more broad terms for specific types of unit system variants.

Unit system variants differ in general based on what they define as "fundamental units." For FPS (Foot-Pound-Second), a customary system, the fundamental unit of length is the foot, the fundamental unit of force is the pound, and fundamental unit of time is the second. For IPS (Inch-Pound-Second), also a customary system, the fundamental unit of length is the inch, where the other fundamental units are the same as FPS.

The Customary system of measurement is desperately annoying. It's the bane of existence for all third graders who are using their newly-honed multiplication and division skills to find how many inches are in 3 feet (36) or how many feet are in 5 yards (15). Length alone can be measured in inches, feet, yards, miles, etc.. Weight can be measured in ounces, pounds, or tons, to name a few options.

Unit Conversions

The metric system was developed with essentially one unit of measurement for each "dimension." That is, length is only measured in some kind of meter. If you want to measure the length of something small, you use a prefix to get something like cm or mm. If you want to measure the length of something large, you use a different prefix to get something like km.

Common unit system variants and their fundamental units are listed below.

Unit System	Variant	Fundamental Units
Customary	FPS Foot-Pound-Second	length – foot (ft) force – pound (lb) time – second (s)
Customary	IPS Inch-Pound-Second	length – inch (in) force – pound (lb) time – second (s)
Metric	mgs meter-gram-second	length – meter (m) mass – gram (g) time – second (s)
Metric	mmgs millimeter-gram-second	length – millimeter (mm) mass – gram (g) time – second (s)
Metric	cgs centimeter-gram-second	length – centimeter (cm) mass – gram (g) time – second (s)
Metric	SI or mks International System of Units or meter-kilogram-second	length – meter (m) mass – kilogram (kg) time – second (s)

Unit Conversion Basics

Since we use two different systems, it is important to know how to convert from one unit system to another. You will occasionally also find yourself converting between units within the same system. One way to think about unit conversions is to use a "multiply-by-one" approach.

Concept 1: Multiplying by 1. Think back to simple fractions. If we have the number ½ and we want the denominator to be 6, we can multiply the top and the bottom by 3. Formally, this would look like this:

$$\frac{1}{2} \cdot \frac{3}{3} = \frac{3}{6}$$

Alternatively, if we had the number 6/15 and wanted the denominator to be 5, we could multiply the top and the bottom by 1/3. (There's a reason we're saying, "multiply the top and bottom by 1/3" instead of "divide the top and bottom by 3." Please bear with us.)

$$\frac{6}{15} \cdot \frac{1/3}{1/3} = \frac{2}{5}$$

There's no reason why we'd have to do this specifically to a fraction. We could just as easily do it to an ordinary number like 15.

$$15 \cdot \frac{3}{3} = \frac{15}{1} \cdot \frac{3}{3} = \frac{45}{3}$$

All these examples work since we're multiplying our original number by a fraction where the top and the bottom numbers are equal (and therefore the fraction itself is equal to 1). We're not changing the actual value of the original number.

Concept 2: Cancelling Like Terms. Now let's do a quick review of how like terms can cancel out in an equation:

$$15x \cdot \frac{y}{3x} = \frac{15x}{1} \cdot \frac{y}{3x} = \frac{15y}{3} = 5y$$

Note that the x's cancel out, and we end up with an answer that is just in terms of y.

Performing a Unit Conversion. We're going to take both of these concepts now and perform an actual unit conversion. Assume we know that there are 3 feet in 1 yard. We want to know how many yards are in 15 feet. (We are converting feet to yards.) Although this is a trivial example, it illustrates a very important concept.

$$15 \text{ ft} \cdot \frac{1 \text{ yd}}{3 \text{ ft}} = \frac{15 \text{ ft}}{1} \cdot \frac{1 \text{ yd}}{3 \text{ ft}} = \frac{15 \text{ yd}}{3} = 5 \text{ yd}$$

Using Concept 1, we're multiplying 15 ft by a fraction where the top and the bottom are actually the same thing. (They're not both the same number, but 1 yd and 3 ft are equal to each other.) Using Concept 2, we're cancelling out the "ft" term from the top and bottom. Finally, we're simplifying 15/3 to get 5.

Performing Another Unit Conversion. Let's try another example. We still know that there are 3 feet in 1 yard. We want to know how many feet are in 12 yards. (We are converting yards to feet.)

$$12 \text{ yd} \cdot \frac{3 \text{ ft}}{1 \text{ yd}} = \frac{12 \text{ yd}}{1} \cdot \frac{3 \text{ ft}}{1 \text{ yd}} = \frac{36 \text{ ft}}{1} = 36 \text{ ft}$$

See how this time for Concept 1, we still multiplied by a fraction where the top and bottom were the same thing, but we "flipped" the fraction from how we did in the previous example. The reason we did this is so that we could use Concept 2, and that the "yd" terms would cancel out.

Using Strange Units. Since sometimes we are working problems where we're learning to use new units for the first time, let's use ones we made up. Assume we know that there are 3 glugs in 2 wakks. We want to know how many glugs are in 5 wakks. (We are converting wakks to glugs.)

$$5 \text{ wakks} \cdot \frac{3 \text{ glugs}}{2 \text{ wakks}} = \frac{5 \text{ wakks}}{1} \cdot \frac{3 \text{ glugs}}{2 \text{ wakks}} = \frac{15 \text{ glugs}}{2} = 7.5 \text{ glugs}$$

The beauty of this method is that if we have any conversion equality, we can easily convert from one unit to another, whether or not the units themselves make any sense.

Here are some useful unit conversion factors (all are exact conversions except the two indicated by ≈):

Customary-Customary	Customary-Metric
12 in = 1 ft	1 in = 2.54 cm
5280 ft = 1 mile	1 mile ≈ 1.609344 km
16 oz = 1 lb	1 lb ≈ 453.59237 grams

In addition to these, don't forget how your unit prefixes come into play here, like 1 km = 10^3 m or 5 cm = $5 \cdot 10^{-2}$ m.

Try these examples on your own:*

   a. 5.8 cm = _____ in
   b. 7 glugs = _____ wakks   (Assume we know that there are 3 glugs in 2 wakks)

Advanced Unit Conversions

There are many ways that unit conversions can start to get tricky.

Converting with multiple conversion factors. Let's say that we only know the conversion factors in the table above, and that we want to convert 12 centimeters to feet. Really, we need to convert centimeters to inches, and then convert inches to feet. The fastest way to do that is to use one conversion statement:

$$12 \text{ cm} \cdot \frac{1 \text{ in}}{2.54 \text{ cm}} \cdot \frac{1 \text{ ft}}{12 \text{ in}} = \frac{12 \text{ cm}}{1} \cdot \frac{1 \text{ in}}{2.54 \text{ cm}} \cdot \frac{1 \text{ ft}}{12 \text{ in}} = \frac{12 \text{ ft}}{30.48} \approx 0.394 \text{ ft}$$

Note we have used the exact same strategy as before, but now both "cm" cancel out and "ft" cancel out.

Converting multiple units. Another problem that can come up is converting something like 625 miles per hour (mph) to kilometers per second.

$$625 \text{ mph} = \frac{625 \text{ miles}}{1 \text{ hour}} \cdot \frac{1.609544 \text{ km}}{1 \text{ mile}} \cdot \frac{1 \text{ hour}}{3600 \text{ s}} = \frac{1005.965 \text{ km}}{3600 \text{ s}} \approx \frac{0.2794347 \text{ km}}{1 \text{ s}}$$

$$\approx 0.279 \text{ km/s}$$

Notice how the first thing we did was to rewrite "625 mph" as 625 miles per 1 hour. This is important so that we know how to position the other unit equivalencies. We also wrote out "miles" instead of writing "m" since "m" is associated with meters, not miles. Also notice that we didn't round off until providing the final answer.

Converting powers of units. The last real confusion can happen when you're converting numbers that represent concepts like areas and volumes. If you have 25 ft², you can't just convert that to in. (The first is an area, while the second is just a length.) Instead, you'd have to convert it to in². Recall that in mathematics, if you square a fraction, you square both the top and bottom of that fraction:

$$\left(\frac{3}{2}\right)^2 = \frac{3^2}{2^2} = \frac{9}{4}$$

Also recall that if you square a product, you square all the terms in that product:

$$(3xy)^2 = 3^2 x^2 y^2 = 9x^2 y^2$$

To use these concepts for unit conversions, let's use the previous example, converting 25 ft² to in².

$$25 \text{ ft}^2 = \frac{25 \text{ ft}^2}{1} \cdot \left(\frac{12 \text{ in}}{1 \text{ ft}}\right)^2 = \frac{25 \text{ ft}^2}{1} \cdot \frac{12^2 \text{ in}^2}{1^2 \text{ ft}^2} = \frac{25 \text{ ft}^2}{1} \cdot \frac{144 \text{ in}^2}{1 \text{ ft}^2} = 3600 \text{ in}^2$$

Here, we still multiplied by 12 in / 1 ft to get the feet to cancel out, but we had to square the conversion term (and therefore all the terms in that equivalency). We didn't just need "ft" to cancel out. We needed "ft-squared" to cancel out.

Pulling it all together. Finally, let's work an example that incorporates all three of these concepts, using "fake" unit systems. Assume there are 3 glugs in 2 wakks and 5 wakks in 9 baks. Convert 3000 glugs³/hour to baks³/s. (Sometimes it really will feel like you're doing complete nonsense anyway, so getting some practice feeling completely lost is good for you!)

$$3000 \frac{\text{glugs}^3}{\text{hour}} = \frac{3000 \text{ glugs}^3}{1 \text{ h}} \cdot \left(\frac{2 \text{ wakks}}{3 \text{ glugs}}\right)^3 \cdot \left(\frac{9 \text{ baks}}{5 \text{ wakks}}\right)^3 \cdot \frac{1 \text{ h}}{3600 \text{ s}}$$

$$= \frac{3000 \text{ glugs}^3}{1 \text{ hour}} \cdot \frac{2^3 \text{ wakks}^3}{3^3 \text{ glugs}^3} \cdot \frac{9^3 \text{ baks}^3}{5^3 \text{ wakks}^3} \cdot \frac{1 \text{ h}}{3600 \text{ s}}$$

$$= \frac{3000 \text{ glugs}^3}{1 \text{ hour}} \cdot \frac{8 \text{ wakks}^3}{27 \text{ glugs}^3} \cdot \frac{729 \text{ baks}^3}{125 \text{ wakks}^3} \cdot \frac{1 \text{ h}}{3600 \text{ s}}$$

$$= \frac{17496000 \text{ baks}^3}{12150000 \text{ s}} \approx \frac{1.477037 \text{ baks}^3}{1 \text{ s}} \approx 1.48 \text{ baks}^3/\text{s}$$

Now see if you can convert these on your own:[†]

   a. 8 m/s = _____ mph
   b. 9 ft²/h = _____ cm²/min
   c. 15 baks/tik = _____ glugs/pic (Assume 9 tiks per 2 pics)

---

[*] a.) 2.28 in b.) 4.67 wakks
[†] a.) 17.9 mph b.) 139 cm²/min c.) 56.25 glugs/pic

Solid Geometry

Modeling, modeling, modeling

Modeling is perhaps THE most important skill to have as an engineer, any type of engineer. One type of modeling is geometric modeling. You should be able to look at an object and simplify its shape in order to approximate things like perimeter, area, surface area, and volume. Geometric formulas for the simplest objects should be in rote memory (you should know them by heart). To that end, think of this section as a miniature geometry class to refresh your memory.

Two-dimensional geometry definitions

Perimeter (P): Distance around an area. The word comes from the Greek "peri" (around) and "meter" (measure), so literally around measure.

Area (A): Quantity describing the two-dimensional size of a surface.

Two-dimensional geometry shapes

Shape	Equations	Image
Rectangle	$P = 2L + 2W$ $A = LW$	
Circle	$P = 2\pi r$ $A = \pi r^2$	

Solid Geometry

Shape	Equations	Image
Triangle	$P = a + b + c$ $A = \frac{1}{2} h\, b$	
Parallelogram	$P = 2b + 2c$ $A = hb$	
Trapezoid	$P = a + b + c + d$ $A = \frac{1}{2}(a + b)\, h$	

Three-dimensional geometry definitions

Surface Area (SA): Area surrounding a defined volume.

Volume (V): Quantity describing the three-dimensional size of an object.

Three-dimensional geometry shapes

Shape	Equations	Image
Rectangular Solid	$SA = 2LH + 2DH + 2DL$ $V = LDH$	

Solid Geometry

Shape	Equations	Image
Cylinder	$SA = 2\pi rh + 2\pi r^2$ $V = \pi r^2 h$	
Sphere	$SA = 4\pi r^2$ $V = \frac{4}{3}\pi r^3$	

Practice using the above equations in the following problems:[*]

    a. What is the volume (in m³) of a cylinder with a height of 80 cm and radius 0.50 m?
    b. What is the area of a circle (in cm²) with a diameter of 1.0 m?
    c. What is the surface area (in m²) of a sphere with a radius of 10 cm?

---

[*] a.) 0.628 m³   b.) 7850 cm²   c.) 0.126 m²

Right Triangles

Right Triangles

Let's look in more detail at a special object, the right triangle. Any triangle with a 90° angle in it is a right triangle. (Can a triangle have more than one 90° angle?*) Right triangles are important because they form the basis of an entire branch of math called Trigonometry. In this section, we will focus on the geometry of the right triangle.

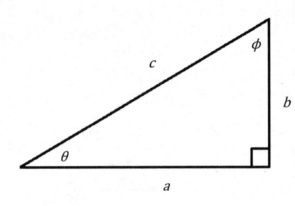

$$c^2 = a^2 + b^2$$

The Pythagorean Theorem relates the length of the hypotenuse to the length of the other two sides.

$$\theta + \phi = 90°$$

This derives from the fact that since the sum of all the angles in any triangle must be 180°, if one angle is 90°, the other two must add up to 90°

There are many "special" right triangles that are important to know. We'll discuss them in the next few subsections.

Pythagorean Triples

There are a few lucky right triangles that all have integer numbers for sides. This is not all that common. If you arbitrarily chose any two lengths for the sides of a right triangle, say 1 and 2, the length of the hypotenuse would be $\sqrt{1^2 + 2^2} = \sqrt{5} \approx 2.236\ ...$, according to the Pythagorean Theorem. However, if you pick another two numbers, 3 and 4, as sides, the length of the hypotenuse would be $\sqrt{3^2 + 4^2} = \sqrt{25} = 5$, which is an integer! Therefore, 3, 4, and 5 together are considered to be a Pythagorean Triple.

A neat consequence is that if 3, 4, 5 is a Pythagorean Triple, then so is 6, 8, 10 (all the elements in the first triple multiplied by 2). Similarly, 3/2, 2, and 5/2 is also a triple. *In fact, any multiple of a Pythagorean Triple is also a Pythagorean Triple.*

While there are methods for generating some Pythagorean Triples, there are only a few that crop up often in engineering textbooks. It's a good idea to have these memorized, so that you're not continually applying the Pythagorean Theorem over and over in your homework!

- 3 – 4 – 5
- 5 – 12 – 13
- 7 – 24 – 25
- 8 – 15 – 17

Again, remember that the largest number has to be the hypotenuse, and that any multiple of these is also a Pythagorean Triple.

Be careful not to fall into the trap of seeing two of these numbers and assuming that it must be a triple! For example, if you have a right triangle with the sides of 3 and 5, the hypotenuse is not 4 – it is actually $\sqrt{3^2 + 5^2} = \sqrt{34} = 5.830$.

The "45-45-90" Triangle

First, consider an isosceles right triangle (a right triangle with two equal sides). We'll label the two equal sides arbitrarily as x. Since the two sides are the same, their opposite angles must be the same, and therefore they must each equal 45°.

If we want to solve for the hypotenuse c, we can do so using the Pythagorean theorem.

$$c^2 = a^2 + b^2$$

$$c^2 = x^2 + x^2$$

$$c^2 = 2x^2$$

$$c = \sqrt{2x^2}$$

$$c = x\sqrt{2}$$

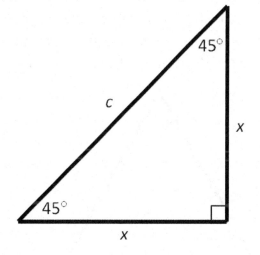

This means that no matter what the sides of the triangle are, the hypotenuse must equal their value times the square root of 2. So, if each of the sides were 5, the hypotenuse must be $5\sqrt{2}$. If the sides were each 100,333,277,545, then the hypotenuse must be 100,333,277,545$\sqrt{2}$.

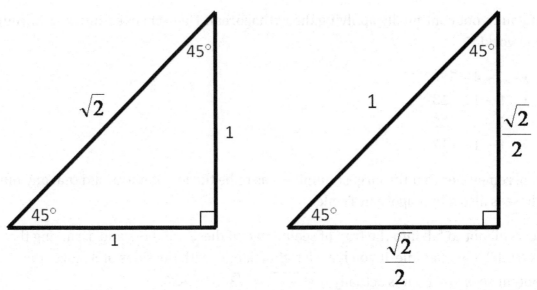

This particular special triangle is usually shown in either of the two ways pictured above. (The second picture is just all the sides in the first picture divided by the square root of two.) The idea is just to show the ratios between all the different sides.

The "30-60-90" Triangle

Now consider an equilateral triangle. All the sides have an equal length of x, and all the angles are 60°. Next, bisect one of the angles with a straight line to get two smaller right triangles, with each having one 30° angle, one 60° angle, and one 90° angle. Each of the smaller legs of the triangles has a length of x.

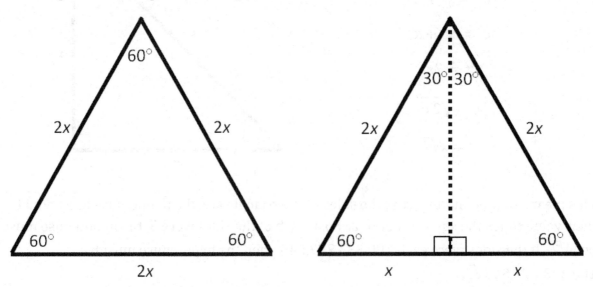

If we want to solve for the remaining side b, we can do so using the Pythagorean theorem.

$$c^2 = a^2 + b^2$$

$$(2x)^2 = x^2 + b^2$$

$$4x^2 = x^2 + b^2$$

$$3x^2 = b^2$$

$$b = \sqrt{3x^2}$$

$$b = x\sqrt{3}$$

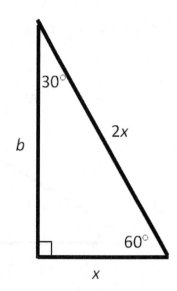

This means that if we have any one of the sides of the triangle, we can find the other sides. If, for example, the side opposite of the 30° angle is 5, then the other leg of the triangle is $5\sqrt{3}$, and the length of the hypotenuse is 10.

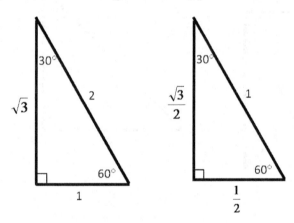

This special triangle is usually shown in either of the two ways pictured to the left. (The second picture is just all the sides in the first picture divided by two.) Again, the idea is just to show the ratios between all the different sides.

Right Triangles

Practice. See if you can figure out the missing sides in the examples below.

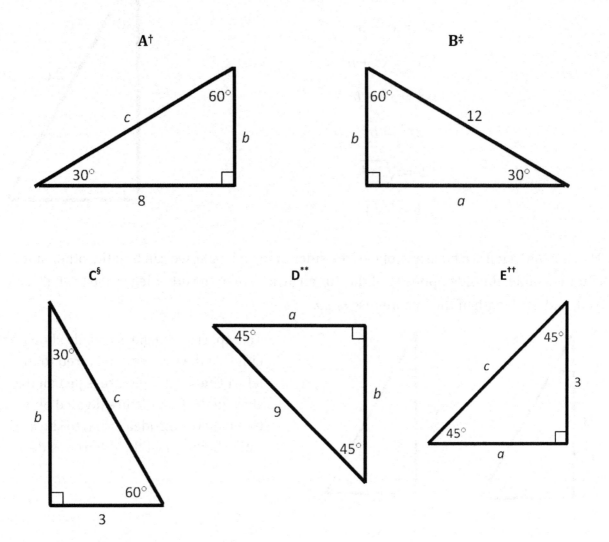

---

[*] No! Triangles can only have 180° total, so the third angle would have to be 0°.

[†] $b = \frac{8}{\sqrt{3}} = \frac{8\sqrt{3}}{3}, c = \frac{16\sqrt{3}}{3}$

[‡] $a = 6\sqrt{3}, b = 6$

[§] $b = 3\sqrt{3}, c = 6$

[**] $a = \frac{9}{\sqrt{2}} = \frac{9\sqrt{2}}{2}, b = \frac{9\sqrt{2}}{2}$

[††] $a = 3, c = 3\sqrt{2}$

Essentials of Trigonometry

Trigonometry Functions

Trigonometry is the study of triangles and the equations associated with them. For the purposes of this book we'll just focus on right triangles. This section should serve as a review for your basic trigonometric equations.

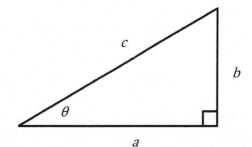

Let's start with some definitions. We're going to concentrate on the angle θ in the drawing to the right.

- The long side c (which is always opposite the right angle) is called the **hypotenuse**.
- The side a is the **adjacent** side to θ
- The side b is the **opposite** side to θ.

The Sine function is defined to be the ratio of the opposite side to the hypotenuse.

$$\sin \theta = \frac{\text{opp}}{\text{hyp}} = \frac{b}{c}$$

The Cosine function is defined to be the ratio of the adjacent side to the hypotenuse.

$$\cos \theta = \frac{\text{adj}}{\text{hyp}} = \frac{a}{c}$$

The Tangent function is defined to be the ratio of the opposite side to the adjacent side.

$$\tan \theta = \frac{\text{opp}}{\text{adj}} = \frac{b}{a}$$

These are definitions. It is from these three definitions that all trigonometric formulas and identities are derived.

Trigonometric functions are useful functions because, among other things, they can help us find unknown sides in right triangles. For example, given the triangle to the right, find c.

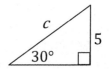

The side opposite to θ is 5, so $a = 5$. The angle is given as 30°. We have the opposite side, the angle, and we are looking for the hypotenuse; therefore, we need to use the sine function.

$$\sin 30° = \frac{5}{c}$$

Solving for c, we obtain the following.

$$c = \frac{5}{\sin 30°} = \frac{5}{0.5} = 10$$

What would the adjacent side be for this triangle?*

Trigonometric functions are also powerful because not only can they be used when you know the angle, but they can also be used to find the angle. We achieve this through the use of "inverse functions," which basically means you're "undoing" a function.*

The inverse functions for sine, cosine, and tangent will also be on your calculator, but may look different. Some calculators show the inverse sine as ASIN, while others use the notation SIN^{-1}. It's also important to note that this isn't "sine to the negative one power," it's the "inverse of the sine function."

Let's see an application of the inverse trigonometric functions. Consider the triangle pictured to the right, and find θ.

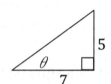

First figure out what information you have about the triangle. The side opposite to θ is 5, so $a = 5$. The side adjacent to θ is 7, so $b = 7$. Since you have the opposite and adjacent sides you need to use the inverse tangent function to calculate the angle.

$$\tan \theta = \frac{5}{7}$$

$$\tan^{-1}(\tan \theta) = \tan^{-1}\left(\frac{5}{7}\right)$$

$$\theta = \tan^{-1}\left(\frac{5}{7}\right) = 35.5°$$

Note: if you get something like 0.620 on your calculator, then that means your calculator is in "radian mode." You will need to review your calculator's manual to determine how to

---

*For example, the cube root function is the inverse of a cube function. The cube root "undoes" the cube.

change the mode to degrees. Or, you can convert from radians to degrees, which we discuss in the next section.

Degrees to Radians

What is the major difference between a degree and a radian? Simply, a degree is an angular measurement while a radian is a measurement of distance along the perimeter of a **unit** circle. Remember the equation for the perimeter of a circle ($P = 2\pi r$). If $r = 1$, then $P = 2\pi$. The 2π radians measurement is the exact distance around the unit circle. In fact, from a geometric perspective this is how 2π is defined.

$$2\pi \text{ radians} \equiv \frac{Perimeter \; of \; a \; circle}{Radius \; of \; a \; circle}$$

Since the perimeter and the radius both have units of distance (meters, for example), the radian must be dimensionless. The radian is the fundamental unit of measurement for a circle, the degree is arbitrary. We could just as easily have defined the degree so that there were 1000° in a circle, but the radian is the fundamental relationship between a circle's radius and its perimeter.

There are 360° in a circle. Those 360° correspond to 2π radians. To write this mathematically, we say:

$$\text{\# Degrees} = \frac{360°}{2\pi \text{ radians}}(\text{\# radians}) = \frac{180°}{\pi \text{ radians}}(\text{\# radians})$$

This means, to convert radians to degrees, multiply by $\pi/180°$.

Example: To how many degrees does $\pi/2$ correspond?

$$\frac{180°}{\pi \text{ radians}}\left(\frac{\pi}{2} \text{ radians}\right) = \frac{180°}{2} = 90°$$

Practice converting from radians to degrees:[†]

a. $\frac{\pi}{3}$ radians
b. $\frac{\pi}{4}$ radians
c. 4 radians

To go from degrees to radians, multiply by the inverse. That is, multiply your degrees by π/180°. Practice converting from degrees to radians:‡

   a. 180°
   b. 225°
   c. 75°

Why can't we just use degrees for everything? Degrees are the units we are most familiar with. Why do we care about radians? It comes down to that fundamental relationship between the radius of a circle and its perimeter.

The arc length of a circle is the distance along the edge of a circle that covers some angle, and is defined as the angle (in radians) times the radius of the circle, or $s = \theta r$. You **have** to change from degrees to radians whenever using angular measurements to get distances.

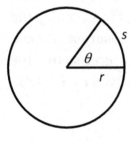

$s = \theta r$

Let's say the radius of the circle is 5 feet and the angle covered is 45°. If you aren't paying attention you might try to tell me that the arc length is 225 ft, but that's bigger than the perimeter of a 5 ft circle [2π(5 ft) = 31.4 ft]. In fact the perimeter of a circle is just the arc length of the entire circle.

The correct answer is 45° (2π/360°) (5 ft) = 3.93 ft, or an eighth of the perimeter. (Think about it, 45° is an eighth of 360°.)

The Unit Circle

The Unit Circle is a special circle that has a radius of one unit. This circle is special because it allows us to have a single picture which gives us the relationship between radians and degrees and allows us to determine **exactly** what sin(θ), cos(θ), and tan(θ) are for a special set of angles.

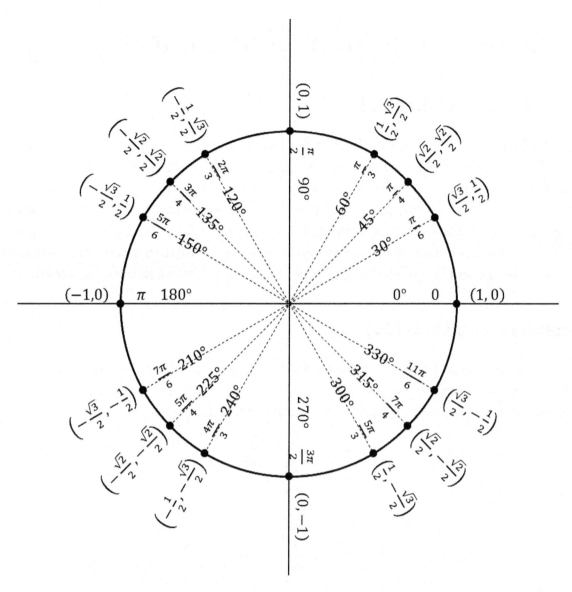

From the inside to the outside, we have the degrees, the radians, and (cos θ, sin θ). For example, cos(7π/6) = -√3/2.

A good engineering student should have all of these values memorized by the time he or she has completed trigonometry. Most engineering professors will assume that you can simplify sin(30°) to 0.5 without a calculator.

---

* $b = 10 \cdot \cos(30°) = 8.66$
† a.) 60°   b.) 45°   c.) 229°
‡ a.) π radians   b.) $\frac{5\pi}{4}$ radians   c.) 1.31 radians

Cartesian and Polar Coordinates

The Importance of Graphing

As an engineering student, one of the first things you're going to come across, probably in Physics class, is the idea that sometimes the numbers work out a lot easier if you look at something in a different way. For example, the equation for graphing a circle of radius 1 in a standard Cartesian coordinate plane is $x^2 + y^2 = 1$. But there is another way of graphing a circle, using something called polar coordinates, where the equation is just $r = 1$. Being able to consider a problem in different coordinate systems can turn an unsolvable problem into a simple one and create mathematical peace out of otherwise impossible situations.

Cartesian Coordinates (2D)

Cartesian coordinates are the "standard" coordinates you're used to. You have probably seen these (at least) in 2-D. To plot a point (x, y) you go x units along the x-axis (the horizontal axis) and y units along the y-axis (the vertical axis).

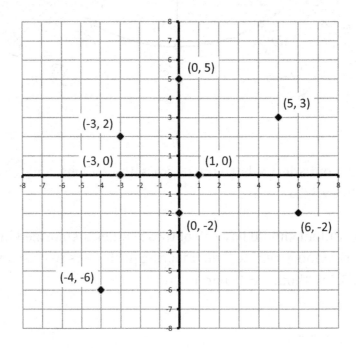

Common familiar shapes plotted in 2D Cartesian coordinates are summarized in the table below:

Shape	General Form of Equation
Line	$Ax + By = C$ or $y = mx + b$
Parabola	$y = ax^2 + bx + c$
Circle	$(x - h)^2 + (y - k)^2 = r^2$
Ellipse	$\frac{x^2}{a^2} + \frac{y^2}{b^2} = 1$ (a and b are arbitrary)
Hyperbola	$\frac{(x-h)^2}{a^2} - \frac{(y-k)^2}{b^2} = 1$ or $\frac{(y-k)^2}{a^2} - \frac{(x-h)^2}{b^2} = 1$

Polar Coordinates

Polar coordinates are based on the idea of arcs and circles. To plot a point (r, θ) you go r units to the right along the primary axis (the one that's in the same place as the x-axis usually is) and then rotate θ degrees (or radians) counterclockwise, maintaining your distance from the origin. Polar coordinates can be plotted in degrees or radians.

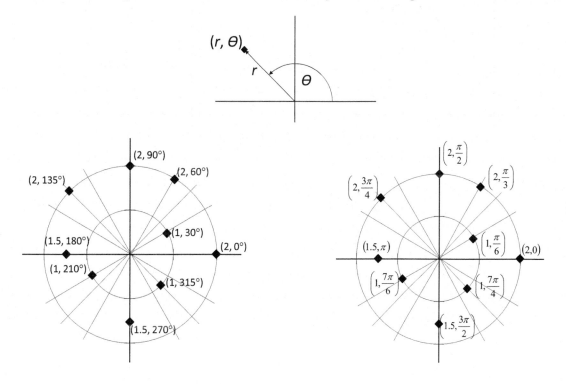

Often polar coordinate "graph paper" uses radial lines extending outward, marking the angles that are multiples of 30° and 45°. Therefore, you'll see radial lines for 30°, 45°, 60°, 120°, 135°, 150°, etc.

An interesting aspect of polar coordinates is that one point can have multiple equivalent coordinates, based on a 360° or 2π circle. However, best practice is usually to provide the θ value between 0° and 360° or 0 and 2π.

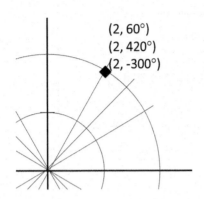

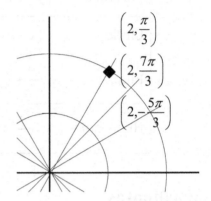

Polar coordinates can also be written with negative r values. However, again, best practice is usually to use $r \geq 0$.

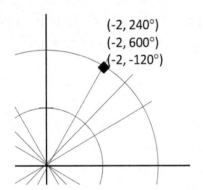

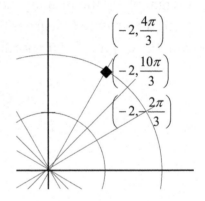

Many interesting figures can be drawn with polar coordinates. One favorite is drawing stars. To draw a five-pointed star, for example, divide the 360° space into 5 parts: 0°, 72°, 144°, 216°, and 288°.

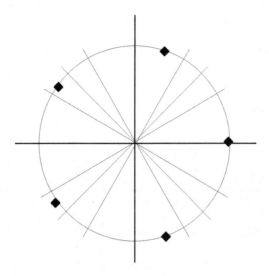

Next, observe that you can create a star from these points by connecting them in the correct order: (1, 0°), (1,144°), (1, 288°), (1, 72°), (1, 216°), and then back to (1, 0°). To make a larger star, make $r > 1$. To create a smaller star, make $0 < r < 1$.

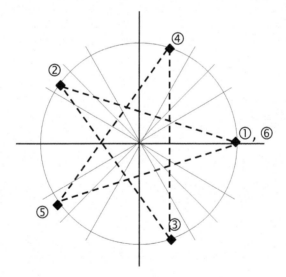

Cartesian and Polar Coordinates

Many unique plots can be created simply with polar coordinates.

Shape	General Form of Equation	Example
Circle	$r = 1, 0 \leq \theta \leq 2\pi$	
Spiral	$r = \theta, 0 \leq \theta \leq 2\pi$	
Flower	$r = \sin(4\theta), 0 \leq \theta \leq 2\pi$	
Heart	$r = 1 - \sin(\theta), 0 \leq \theta \leq 2\pi$	

Converting Between Cartesian and Polar Coordinates

You can convert between polar and Cartesian coordinates using trigonometric functions. If you love to memorize formulas and turn your brain off for later to just chug through problems, here you go:

Cartesian to Polar	Polar to Cartesian
$r = \sqrt{x^2 + y^2}$ $\theta = \begin{cases} 0 & x = 0 \text{ and } y = 0 \\ \sin^{-1}\frac{y}{r} & x \geq 0 \\ -\sin^{-1}\frac{y}{r} + \pi & x < 0 \end{cases}$	$x = r \cos \theta$ $y = r \sin \theta$

The adjustment to θ is due to the fact that the trigonometric functions are not one-to-one. That is, the sine of 30° is 0.5, but the sine of 150° is also 0.5. To distinguish between the two options, we must use the piecewise function given above.

As an alternative to just memorizing more formulas, you can use trigonometry to "figure it out" each time you need it. (This method is probably the most effective – at some point, your brain is going to need some of this space back, and in theory you should have all this trig stuff memorized by now anyway.)

Consider the example again where we plotted (2, 60°). To convert these coordinates from polar to Cartesian, imagine a triangle with one point at the origin, another point on the x-axis, and then a line running perpendicularly from there straight up to the polar point. Next, use trigonometry to solve for the missing x and y values, getting that x = 1 and y = 1.73. This means that the Cartesian equivalent of the polar point (2, 60°) is (1, 1.73).

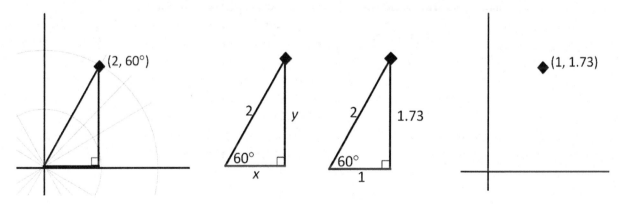

Now, let's go from Cartesian to polar. We'll use a slightly more complicated example. First, recognize that the two Cartesian points represent the two sides of a right triangle. We can use trigonometry to find the missing values in that right triangle.

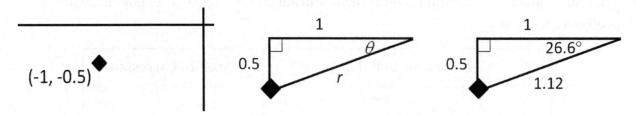

Now, before you declare victory, be sure and look at what you've actually calculated.

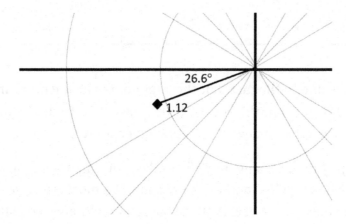

Notice that the 26.6° isn't the right angle for the polar coordinate placement. In fact, we have to add 180° to that value to make it correct. As a final check, we can see where the point is and make sure it at least "looks" like it's in the right place.

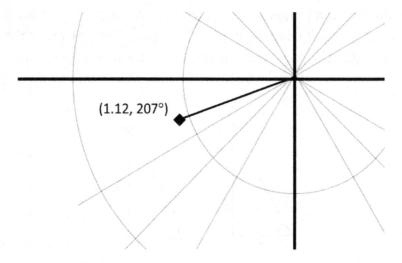

Basics of Matrices

What is a matrix?

Simply, a matrix is a rectangular array of numbers, usually represented algebraically by a capital letter.

$$A = \begin{bmatrix} 3 & 2 \\ -9 & 7 \\ 14 & 5 \end{bmatrix}, \quad B = \begin{bmatrix} 6 \\ 28 \\ -9 \end{bmatrix}, \quad C = [-9.7 \quad 5.6 \quad 3.1],$$

$$D = \begin{bmatrix} 6 & -9.98 & 8+5i \\ e^2 & \cos 45° & \sqrt{x} \\ 75 & \ln 5 & -98 \end{bmatrix}, \quad \text{or } E = [5]$$

All of the above are matrices. The elements of a matrix can be any mathematical expression. A matrix with m rows and n columns is said to be an $m \times n$ matrix.

Technically, a single value like 5 is a 1×1 matrix, so you've been using matrices all your life. The matrices above have dimensions of 3×2, 3×1, 1×3, 3×3, and 1×1, respectively.

An $m \times 1$ matrix is called a **row vector**, while a $1 \times n$ matrix is a **column vector**.

$$A = [5.6 \quad 7.2 \quad 9.3], \quad B = \begin{bmatrix} 16 \\ -2 \\ 5 \end{bmatrix}$$

A 1×1 matrix is called a **scalar**.

$$C = [-2.36] = -2.36$$

Can you determine the sizes ($m \times n$) of the following matrices?*

$$A = \begin{bmatrix} 6 & 2 \\ 18 & 10 \\ 1 & 4 \end{bmatrix} \qquad C = \begin{bmatrix} 4 & -7 & 9 & 8 \\ 7 & 3 & -2 & 3 \\ -1 & 0 & 16 & 5 \\ 3 & 87 & 1 & 14 \end{bmatrix}$$

$$B = \begin{bmatrix} 7 \\ 8 \\ 99 \end{bmatrix} \qquad D = [67 \quad 8.9 \quad 13 \quad 45]$$

Let's say we have the matrix

$$C = \begin{bmatrix} 4 & -7 & 9 & 8 \\ 7 & 3 & -2 & 3 \\ -1 & 0 & 16 & 5 \\ 3 & 87 & 1 & 14 \end{bmatrix}$$

We might want to reference a single element from that matrix; say the 87. It is in the 4th row from the top and the 2nd column from the left. We normally reference elements with the row number first and the column number second. Instead of writing the sentence, "The number 87 is in the 4th row and 2nd column of the matrix C," we can use a number pair to reference an element.

$$C(4,2) = 87$$

means "the 4th row and 2nd column element of C is 87."

Practice finding elements:[†]

 a. $C(4,1)$
 b. $C(3,3)$
 c. $C(1,2)$

It is more usual to use something called index notation for a matrix, explained with the following picture.

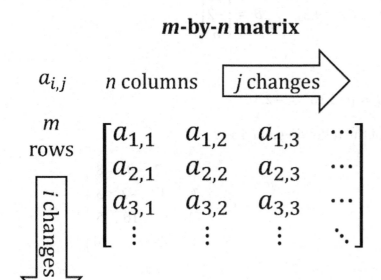

If your matrix is represented by the capital letter A, it is common practice to use its lowercase counterpart a to represent the matrix's elements.

The individual elements of a matrix are then referenced by turning their row and column position into subscripts. So the element in the row i and column j position would be written a_{ij}.

Let's say we have the matrix

$$C = \begin{bmatrix} 4 & 7 & 9 & 8 \\ 7 & 3 & -2 & 3 \\ -1 & 0 & 16 & 5 \\ 3 & 87 & 1 & 14 \end{bmatrix}$$

then $c_{4,2} = 87$.

Practice finding these matrix elements:[‡]

a. $c_{3,4}$
b. $c_{4,1}$
c. $c_{2,2}$

Why use matrices?

Here is a list of only some of the things you can use matrices for:

- Solving systems of linear equations
- Representing linear transformations
- Solving ordinary differential equations
- Probability theory and statistics
- Symmetries and transformations in physics (mechanics)
- Quantum mechanics
- Geometric optics
- Circuit design
- Computer programming
- Computer graphics

Don't worry if you don't know what some of the things in the list are, if you continue your study of engineering you will come across some of them eventually. The point is that matrix math is extremely important for anyone modeling physical systems.

Matrix Math (For all types of matrices)

For the sake of sanity we're mostly going to use 2×2 matrices in the following section with a few 3×3 matrices thrown into the mix. The following rules work for all types of matrices.

Basics of Matrices

Matrix addition: Add corresponding elements together directly.

$$A + B = \begin{bmatrix} a_{1,1} & a_{1,2} \\ a_{2,1} & a_{2,2} \end{bmatrix} + \begin{bmatrix} b_{1,1} & b_{1,2} \\ b_{2,1} & b_{2,2} \end{bmatrix} = \begin{bmatrix} a_{1,1} + b_{1,1} & a_{1,2} + b_{1,2} \\ a_{2,1} + b_{2,1} & a_{2,2} + b_{2,2} \end{bmatrix} = C$$

If we write this equation element-wise:

$$c_{i,j} = a_{i,j} + b_{i,j}$$

Both $A + B = C$ and $c_{i,j} = a_{i,j} + b_{i,j}$ are basically the same equation. Matrix addition is commutative: $A + B = B + A$. This may seem obvious and it is not difficult to prove, but there are matrix operations that are not commutative, like matrix multiplication which we will discuss in a moment. Also note, to add matrices they MUST BE the SAME SIZE.

Example:

$$\begin{bmatrix} 5 & 3 \\ 2 & -1 \end{bmatrix} + \begin{bmatrix} 10 & -1 \\ 67 & 18 \end{bmatrix} = \begin{bmatrix} 5+10 & 3-1 \\ 2+67 & -1+18 \end{bmatrix} = \begin{bmatrix} 15 & 2 \\ 69 & 17 \end{bmatrix}$$

Practice with the following:§

a. $\begin{bmatrix} 1 & -4 \\ 2 & 5 \end{bmatrix} + \begin{bmatrix} 5 & 8 \\ 0 & -1 \end{bmatrix} = ?$

b. $\begin{bmatrix} 10 & 14 & 5 \\ 6 & -2 & 3 \\ 4 & 9 & 7 \end{bmatrix} + \begin{bmatrix} 5 & 8 \\ 0 & -1 \end{bmatrix} = ?$

c. $\begin{bmatrix} 10 & 14 & 5 \\ 6 & -2 & 3 \\ 4 & 9 & 7 \end{bmatrix} + \begin{bmatrix} -5 & 2 & 6 \\ 0 & 7 & 1 \\ 18 & 56 & 3 \end{bmatrix} = ?$

Multiplication by a scalar (1×1 matrix): Multiply each element by the scalar.

$$cA = c \begin{bmatrix} a_{1,1} & a_{1,2} \\ a_{2,1} & a_{2,2} \end{bmatrix} = \begin{bmatrix} ca_{1,1} & ca_{1,2} \\ ca_{2,1} & ca_{2,2} \end{bmatrix} = D$$

$$\text{or } d_{i,j} = ca_{i,j}$$

Basics of Matrices

We can combine addition and scalar multiplication. For example, let

$$A = \begin{bmatrix} 6 & -2 & 2 \\ 4 & 0 & 3 \\ 1 & -7 & 1 \end{bmatrix}, \quad B = \begin{bmatrix} 3 & -3 & 0 \\ 5 & 1 & 4 \\ -1 & 2 & 2 \end{bmatrix}, \quad c = 5.$$

What is $D = A + cB$?

$$D = \begin{bmatrix} 6 & -2 & 2 \\ 4 & 0 & 3 \\ 1 & -7 & 1 \end{bmatrix} + 5\begin{bmatrix} 3 & -3 & 0 \\ 5 & 1 & 4 \\ -1 & 2 & 2 \end{bmatrix} = \begin{bmatrix} 6 & -2 & 2 \\ 4 & 0 & 3 \\ 1 & -7 & 1 \end{bmatrix} + \begin{bmatrix} 15 & -15 & 0 \\ 25 & 5 & 20 \\ -5 & 10 & 10 \end{bmatrix} = \begin{bmatrix} 21 & -17 & 2 \\ 29 & 5 & 23 \\ -4 & 3 & 11 \end{bmatrix}.$$

Matrix Multiplication: This gets tricky.

Matrix multiplication is much more complicated than scalar multiplication. In reality, you have to first define what you mean by multiplication and that's not as easy as it sounds. In fact the definition of multiplication is best left to a linear algebra class. There are actually a few ways to do matrix multiplication, but I will focus on the most common here.

The product AB of two matrices is defined only if the number of columns of A is equal to the number of rows of B. Let's say A is 2×3, we can multiply A times B if B is 3×2, 3×5, or even 3×3000, basically 3×anything would work. If A is $m \times r$, B is $r \times n$, and $C = AB$, then C will be $m \times n$.

If $AB = C$, then

$$c_{i,j} = \sum_{k=1}^{r} a_{i,k} b_{k,j}$$

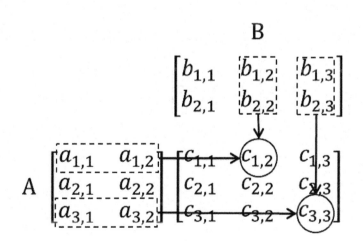

The picture to the left shows $A(3\times2)$ times $B(2\times3)$ equals $C(3\times3)$. The element for the top circle would be

$$c_{1,2} = a_{1,1}b_{1,2} + a_{1,2}b_{2,2}$$

And the element in the lower circle would be

$$c_{3,3} = a_{3,1}b_{1,3} + a_{3,2}b_{2,3}$$

For every element in C multiplication AND summation is required.

Basics of Matrices

Example: Let

$$A = \begin{bmatrix} 6 & -2 & 2 \\ 4 & 0 & 3 \\ 1 & -7 & 1 \end{bmatrix}, \quad B = \begin{bmatrix} 3 & -3 & 0 \\ 5 & 1 & 4 \\ -1 & 2 & 2 \end{bmatrix}.$$

What is $D = AB$?

$$D = \begin{bmatrix} 6 \cdot 3 + (-2) \cdot 5 + 2 \cdot (-1) & 6 \cdot (-3) + (-2) \cdot 1 + 2 \cdot 2 & 6 \cdot 0 + (-2) \cdot 4 + 2 \cdot 2 \\ 4 \cdot 3 + 0 \cdot 5 + 3 \cdot (-1) & 4 \cdot (-3) + 0 \cdot 1 + 3 \cdot 2 & 4 \cdot 0 + 0 \cdot 4 + 3 \cdot 2 \\ 1 \cdot 3 + (-7) \cdot 5 + 1 \cdot (-1) & 1 \cdot (-3) + (-7) \cdot 1 + 1 \cdot 2 & 1 \cdot 0 + (-7) \cdot 4 + 1 \cdot 2 \end{bmatrix}$$

$$= \begin{bmatrix} 18 - 10 - 2 & -18 - 2 + 4 & 0 - 8 + 4 \\ 12 + 0 - 3 & -12 + 0 + 6 & 0 + 0 + 6 \\ 3 - 35 - 1 & -3 - 7 + 2 & 0 - 28 + 2 \end{bmatrix} = \begin{bmatrix} 6 & -16 & -4 \\ 9 & -6 & 6 \\ -33 & -8 & -26 \end{bmatrix}.$$

Matrix multiplication is **not** commutative, $AB \neq BA$.

Matrix Transpose: Flip the rows and columns.

The transpose of a matrix is formed by turning rows into columns and vice versa. The transpose is denoted with a superscript T. Let

$$A = \begin{bmatrix} a_{1,1} & a_{1,2} \\ a_{2,1} & a_{2,2} \end{bmatrix} \text{ and } B = A^T$$

Then

$$B = \begin{bmatrix} a_{1,1} & a_{2,1} \\ a_{1,2} & a_{2,2} \end{bmatrix} \text{ or } b_{i,j} = a_{j,i}$$

Notice that the main diagonal does not change; the transpose is a "flip" around the main diagonal.

Example:

$$A = \begin{bmatrix} 6 & -2 & 2 \\ 4 & 0 & 3 \\ 1 & -7 & 1 \end{bmatrix} \quad A^T = \begin{bmatrix} 6 & 4 & 1 \\ -2 & 0 & -7 \\ 2 & 3 & 1 \end{bmatrix}$$

Matrix Math (Only for square matrices)

The following matrix math only works (or is defined) for square matrices, matrices that are $n \times n$ (same number of rows and columns).

The Identity Matrix:

The identity matrix (I) is like the 1 in regular math. You can multiply I times anything and just get the original matrix back. That is, $AI = A$ regardless of A's matrix properties, although A and I still both have to be square and have the same dimensions. The 1×1 identity matrix I is 1. Other I matrices include

$$\begin{bmatrix} 1 & 0 \\ 0 & 1 \end{bmatrix} \text{ and } \begin{bmatrix} 1 & 0 & 0 \\ 0 & 1 & 0 \\ 0 & 0 & 1 \end{bmatrix}$$

The identity matrix has ones along the main diagonal and zeros everywhere else.

Powers of Matrices:

First we have to remember what raising a value to a power means. For example, x^2 is x times x. So, $A^2 = AA$. The only way to do this multiplication is if the number of rows of A equals the number of columns of A. Thus, A must be square for A^2 to have any meaning.

$$A^2 = AA, \quad A^3 = (AA)A = A^2A, \quad A^n = A^{n-1}A$$

Matrix Inverse:

Only square matrices can have inverses but not all square matrices do have inverses. We denote the inverse of a matrix with a superscript -1. If the inverse of A exists it is the matrix such that

$$AA^{-1} = I \quad \text{(and also } A^{-1}A = I\text{)}$$

That is, a matrix times its inverse gives the identity matrix. Which matrices have inverses? To determine this, we have to talk about something called the determinant.

Determinant of a Matrix:

The determinant is actually difficult to write as a formula in general without relying on indices and linear algebra, but we can quickly write a formula for both the 2×2 matrix and the 3×3 matrix.

$$\text{If } A = \begin{bmatrix} a & b \\ c & d \end{bmatrix} \text{ then } \det(A) = ad - bc.$$

If $A = \begin{bmatrix} a & b & c \\ d & e & f \\ g & h & i \end{bmatrix}$ then $\det(A) = aei + bfg + cdh - ceg - afh - bdi$

You don't have to actually memorize the above equation. There's actually a more visual way to consider it:

$+ \text{ parts} = \begin{bmatrix} a & b & c & a & b \\ d & e & f & d & e \\ g & h & i & g & h \end{bmatrix} = aei + bfg + cdh$

$- \text{ parts} = \begin{bmatrix} a & b & c & a & b \\ d & e & f & d & e \\ g & h & i & g & h \end{bmatrix} = -ceg - afh - bdi$

So the determinant is the + parts plus the − parts. (This only works for a 3 x 3 matrix.) Doing determinants for larger matrices is a pain, so we won't talk about it here.

Notice that the determinant is just a number, not a matrix. Here comes a really important theory.

A matrix has an inverse if and only if its determinant is nonzero.

Example: Let

$$A = \begin{bmatrix} 6 & -2 & 2 \\ 4 & 0 & 3 \\ 1 & -7 & 1 \end{bmatrix}.$$

What is $\det(A)$?

$$\det(A) = 6 \cdot 0 \cdot 1 + (-2) \cdot 3 \cdot 1 + 2 \cdot 4 \cdot (-7) - 2 \cdot 0 \cdot 1 - (-2) \cdot 4 \cdot 1 - 6 \cdot 3 \cdot (-7)$$
$$= 0 - 6 - 56 - 0 + 8 + 126 = 72$$

Thus A has an inverse. (I don't know what it is but I know it has one! Actually it is hard to calculate an inverse for any general matrix, and the process is beyond the scope of this book.)

Matrix Application (Systems of Linear Equations)

As mentioned above, there are many applications for matrices, such as solving systems of linear equations, representing linear transformations, solving ordinary differential equations, etc. We will focus here on systems of linear equations.

Let us consider a system of three equations:

$$c_1 x_1 + c_2 x_2 + c_3 x_3 = b_1$$

$$c_4 x_1 + c_5 x_2 + c_6 x_3 = b_2$$

$$c_7 x_1 + c_8 x_2 + c_9 x_3 = b_3$$

This system of equations can be represented as one matrix equation. Let

$$A = \begin{bmatrix} c_1 & c_2 & c_3 \\ c_4 & c_5 & c_6 \\ c_7 & c_8 & c_9 \end{bmatrix}, \quad X = \begin{bmatrix} x_1 \\ x_2 \\ x_3 \end{bmatrix}, \quad B = \begin{bmatrix} b_1 \\ b_2 \\ b_3 \end{bmatrix}$$

Then

$$AX = B \text{ or } \begin{bmatrix} c_1 x_1 + c_2 x_2 + c_3 x_3 \\ c_4 x_1 + c_5 x_2 + c_6 x_3 \\ c_7 x_1 + c_8 x_2 + c_9 x_3 \end{bmatrix} = \begin{bmatrix} b_1 \\ b_2 \\ b_3 \end{bmatrix}$$

To solve a system of equations, all we need to do is solve ONE matrix equation. We need to solve

$$AX = B$$

for X. If we multiply both sides of that equation with the inverse of A, we get

$$A^{-1} AX = A^{-1} B$$

$$IX = A^{-1} B$$

$$X = A^{-1} B$$

So, it is always possible to find a solution to any system of equations *as long as the inverse of A exists*. We showed earlier in this section that if $\det(A) \neq 0$ then the matrix has an inverse. Therefore, if $\det(A) \neq 0$ the system of equations has a unique solution. This idea applies to any size matrix; thus, any size system of equations. For example if you have ten equations and ten unknowns, find A and take its determinant. If $\det(A) \neq 0$, the system has a solution. This is a quick and easy way to check if a system is solvable, before you have gone through the effort of trying to solve it.

To solve a system of equations, normally we do not want to go through the process of solving for one variable from one equation, plugging it into another of the equations, solving for the next variable from the new equation, and so on. This is relatively easy for a system with only two equations and two unknowns, and is doable for a system with three equations and three unknowns; but quickly becomes tedious and time consuming with more equations and more unknowns. It is also tedious to find the inverse of a matrix for any matrix bigger than a 3×3. So, normally you use something called the reduced row echelon form to find the solution to your system of equations. We are going to skip a lot of the linear algebra involved in detailing why this works, as it is beyond the scope of this book, and just show you the process.

Start with the system of three equations

$$c_1 x_1 + c_2 x_2 + c_3 x_3 = b_1$$

$$c_4 x_1 + c_5 x_2 + c_6 x_3 = b_2$$

$$c_7 x_1 + c_8 x_2 + c_9 x_3 = b_3$$

Represented by

$$A = \begin{bmatrix} c_1 & c_2 & c_3 \\ c_4 & c_5 & c_6 \\ c_7 & c_8 & c_9 \end{bmatrix}, \quad X = \begin{bmatrix} x_1 \\ x_2 \\ x_3 \end{bmatrix}, \quad B = \begin{bmatrix} b_1 \\ b_2 \\ b_3 \end{bmatrix}$$

We are going to use a matrix which combines A and B

$$D = \begin{bmatrix} c_1 & c_2 & c_3 & b_1 \\ c_4 & c_5 & c_6 & b_2 \\ c_7 & c_8 & c_9 & b_3 \end{bmatrix}$$

We can use elementary row operations to get this matrix into its reduced row echelon form. Elementary row operations include multiplying a row by a scalar and adding and subtracting one row from another. One important theorem from linear algebra states that using elementary row operations on a matrix representing a system of equations doesn't change the solutions to the system of equations.

In reduced row echelon form the matrix D will now look like

$$D' = \begin{bmatrix} 1 & 0 & 0 & b'_1 \\ 0 & 1 & 0 & b'_2 \\ 0 & 0 & 1 & b'_3 \end{bmatrix}$$

The accent mark is a reminder that D' is not the same matrix as D and the b' numbers will be different from the original b numbers.

Now split D' back into an A' and a B' such that

$$A' = \begin{bmatrix} 1 & 0 & 0 \\ 0 & 1 & 0 \\ 0 & 0 & 1 \end{bmatrix}, \quad X = \begin{bmatrix} x_1 \\ x_2 \\ x_3 \end{bmatrix}, \quad B' = \begin{bmatrix} b'_1 \\ b'_2 \\ b'_3 \end{bmatrix}$$

These matrices still represent the original system of equations, but now

$$A'X = B'$$

Gives us

$$x_1 = b'_1$$

$$x_2 = b'_2$$

$$x_3 = b'_3$$

And we are done! We have the solution to our system of equations.

Finding the reduced row echelon form for a matrix by hand can also be tedious and time consuming; luckily, most graphing calculators have functions which will do most of the work for us.

Let

$$8x_1 - 5x_2 + 2x_3 - 1 = 0$$

$$x_2 + 6x_3 = 5$$

$$-3x_3 + 2x_1 = 8$$

Then

$$A = \begin{bmatrix} 8 & -5 & 2 \\ 0 & 1 & 6 \\ 2 & 0 & -3 \end{bmatrix}, \quad X = \begin{bmatrix} x_1 \\ x_2 \\ x_3 \end{bmatrix}, \quad B = \begin{bmatrix} 1 \\ 5 \\ 8 \end{bmatrix}$$

$$D = \begin{bmatrix} 8 & -5 & 2 & 1 \\ 0 & 1 & 6 & 5 \\ 2 & 0 & -3 & 8 \end{bmatrix}$$

Notice that we've had to do some work to get the system of equations into "standard form." For example, on the first equation, we needed to add one to both sides of the equation to get $8x_1 - 5x_2 + 2x_3 = 1$. We've also put in zeros in places where the original equations were "missing" variables. The second equation $x_2 + 6x_3 = 5$, for example, doesn't seem to have a value for x_1. We know, however, that it really does, it's just that its coefficient is zero. That is, $0x_1 + x_2 + 6x_3 = 5$. Therefore, when we create matrix A, we must put a "placeholder" in the spot for x_1.

Also notice we had to rearrange the third equation so that all the x's appeared in order. To make the third equation usable $-3x_3 + 2x_1 = 8$, had to be rewritten as $2x_1 - 3x_3 = 8$ (or $2x_1 + 0x_2 - 3x_3 = 8$) and then its coefficients placed in A.

Basics of Matrices

The matrix D is the matrix you want to input into the graphing calculator.

1. On the TI-84, hit 2ND and then MATRIX.
2. Arrow right to EDIT, and down to [D].
3. Hit ENTER.
4. Enter 3×4 for the matrix size, and then input the individual entries in the matrix.
5. Hit 2ND and QUIT.
6. Hit 2ND and MATRIX again.
7. Arrow right to MATH and down to RREF (.
8. Hit ENTER.
9. Hit 2ND and MATRIX again.
10. Arrow down to [D] and hit ENTER.
11. Add the closing parenthesis and hit ENTER.

You should see this for D'

$$D' = \begin{bmatrix} 1 & 0 & 0 & 3.795454545 \\ 0 & 1 & 0 & 5.818181818 \\ 0 & 0 & 1 & -0.1363636364 \end{bmatrix}$$

Note that we see the identity matrix on the left, so that means we have a solution, and the solutions for this system of equations are

$$x_1 = 3.795454545$$

$$x_2 = 5.818181818$$

$$x_3 = -0.1363636364$$

to the correct number of significant figures.

If you ever see an answer that looks like this,

$$D' = \begin{bmatrix} 1 & 0 & 0 & 3.795454545 \\ 0 & 1 & 1 & 5.818181818 \\ 0 & 0 & 0 & -0.1363636364 \end{bmatrix}$$

you should immediately recognize that you don't have an identity matrix on the left (see that there are two ones in the second row), and therefore, there would be no solution to the system of equations that you entered.

Basics of Matrices

As a side note, sometimes you will get solutions like this:

$$x_1 = 2.45\text{E-}15$$

$$x_2 = 5.6$$

$$x_3 = 6.7$$

Note that in calculator speak, $x_1 = 2.45\text{E-}15$ really means $x_1 = 2.45 \times 10^{-15}$. Since the calculator uses a numerical method to get the rref, the small number represents a numerical error. The solution is actually:

$$x_1 = 0$$

$$x_2 = 5.6$$

$$x_3 = 6.7$$

Try to find the solution for the following:[**]

$$5x_1 + 10x_2 - x_3 = 5$$

$$-2x_1 + 2x_2 + 4x_3 = 0$$

$$x_1 + x_2 - 2x_3 = 4$$

---

[*] a.) 3 x 2   b.) 3 x 1   c.) 4 x 4   d.) 1 x 4
[†] a.) 3   b.) 16   c.) -7
[‡] a.) 5   b.) 3   c.) 3
[§] a.) $\begin{bmatrix} 6 & 4 \\ 2 & 4 \end{bmatrix}$ b.) No answer!   c.) $\begin{bmatrix} 5 & 16 & 11 \\ 6 & 5 & 4 \\ 22 & 65 & 10 \end{bmatrix}$
[**] $x_1 = -3.56$, $x_2 = 2.00$, $x_3 = -2.78$

Lab Measurements and Error Analysis

Introduction

All measurements have uncertainties. There is no such thing as a perfectly accurate measurement. Your ruler, voltmeter, or thermometer only has a finite number of divisions. They all have a smallest value which they can measure.

Science and English use the same words, but often words used in science have a different connotation or a more precise definition. Words like **precision** and **accuracy** are synonyms in everyday usage, but scientifically they are used to describe completely different things. In everyday usage **error** carries a negative connotation, like a mistake or a problem. In scientific language, the term **error** simply denotes uncertainty with no negative connotation. Uncertainty is NOT a bad thing; it's a physical fact like gravity or conservation of energy.

Precision

Precision – the repeatability of a measurement using a specific instrument.

You can think of precision as the smallest measurement an instrument can give. The best way to examine this is by example.

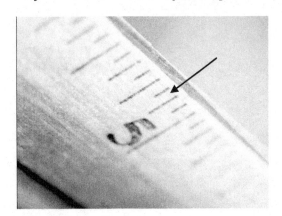

Take the ruler to the left. The smallest division on this ruler is $\frac{1}{16}$th of an inch. You can estimate with some accuracy to about half of that value, so the precision for this ruler is $\frac{1}{32}$nd of an inch.

Let's say we measure something to the arrow in the image. Then the measurement would be 4 and $\frac{13}{16}$ths of an inch with an uncertainty of $\frac{1}{32}$nd. You would report this as:

$$4\frac{13}{16} \pm \frac{1}{32} \text{ inches}$$

If you are trying to write this in decimal form, you only keep ONE significant figure for the precision. Thus, the above measurement would be reported as 4.81±0.03 inches.

This tells someone reading your data that the object measured was, to your best estimate, 4.81 inches long, and that you are relatively certain that its length lies between 4.78 inches and 4.84 inches.

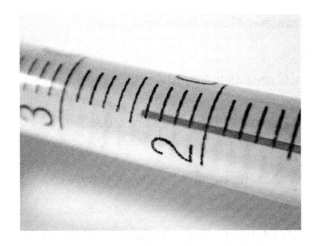

Try another example. Take the thermometer to the left. The smallest division on this ruler is 1 °C. As above, you can estimate with some accuracy to about half of that value, so the precision for this thermometer is 0.5 °C.

The fluid in the thermometer is slightly above the 24 °C mark, but not halfway to the 25 °C mark. So the measurement here would be 24.0 ± 0.5 °C.

If somebody else came along and said, I think the fluid is closer to the halfway mark then they would report the temperature as 24.5 ± 0.5 °C. BOTH results are absolutely correct, this is one of the reasons we take multiple measurements and average the results.

Let's try one more example.

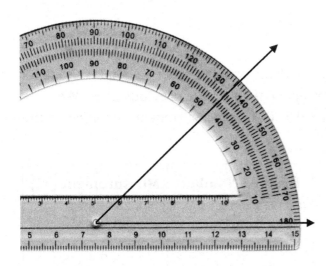

Take the protractor to the left. The smallest division on this ruler is 1°. As above, you can estimate with some accuracy to about half of that value, so the precision for this protractor is 0.5°.

The line representing the angle is slightly below the 45° mark, but not halfway to the 44° mark. So the measurement here would be 45.0 ± 0.5°.

If somebody else came along and said, I think the line is closer to the halfway mark then they would report the angle as 44.5 ± 0.5°. Again, BOTH results are absolutely correct.

Accuracy

Accuracy – how close we are to the "true" value with our measurement.

True has quotes around it because all physical parameters have some uncertainty associated with them.

For example, let's say that the thermometer above is measuring room temperature. Say we have also a calibrated thermocouple (a much more precise device) that shows us the temperature in the room is 24.32 ± 0.01 °C. Calibrated means that we know the thermocouple gives accurate data.

Is our thermometer accurate? Yes, 24.32 °C falls within the range of values given by the measurements made by the thermometer.

Let's say instead that the thermocouple reads 22.45 ± 0.01 °C. This would mean that our thermometer was NOT accurate and/or that there was something wrong with our measurements.

The above shows qualitative ways of looking at accuracy. To discuss accuracy quantitatively we need to talk about error analysis.

Error Analysis

The best way to look at error analysis is to start with an example. Let's say that we took ten measurements of room temperature with the thermometer on the previous page. We know now that this data should look like ##.# ± 0.5 °C. Our measurements are given in the table below.

Reading Number	Measurement (°C)	Reading Number	Measurement (°C)
1	24.0 ± 0.5	6	24.0 ± 0.5
2	23.5 ± 0.5	7	25.0 ± 0.5
3	24.0 ± 0.5	8	23.0 ± 0.5
4	22.0 ± 0.5	9	23.5 ± 0.5
5	24.5 ± 0.5	10	24.5 ± 0.5

These are our data points, all with a precision of ±0.5 °C. Notice that some of the values are separated by more than our precision. Does this mean that the data is bad? NO. The temperature in the room could easily be fluctuating by this much over the time period that the data was taken.

Both the measurement estimates we make based on the precision of the instrument and the fluctuation of the temperature in the room are sources of uncertainty. These uncertainties have to be combined to find an overall value of uncertainty for our experiment.

In order to deal with combinations of uncertainties quantitatively, the uncertainties must be **random**. By random, we mean that the measured values need to fall symmetrically around a central value, usually the mean or average value.

If the uncertainties are not random they are **systematic**, that is, they are errors caused by the way we are running the experiment, and must be removed before the experiment takes place. For example, if the thermometer consistently reads two degrees below the "true" value, the thermometer needs to be replaced.

Often there are multiple sources of random errors in an experiment, and the experimenter does not know the source of the errors or have any way to separate them. Statistical analysis of data accounts for this problem. Let's define some statistical terms and then discuss how to use them.

Mean – The average value of a set of data points.

You are probably already familiar with this. The formula is:

$$\bar{x} = x_{mean} = \frac{\sum_{i=1}^{N} x_i}{N}$$

where x is the value being measured, N is the total number of measurements, and the sum is over all measurements. From our data above (without the uncertainty from precision):

$$24.0, 23.5, 24.0, 22.0, 24.5, 24.0, 25.0, 23.0, 23.5, 24.5$$

$$T_{mean} = \frac{\sum_{i=1}^{10} T_i}{10} = \frac{24.0 + 23.5 + 24.0 + 22.0 + 24.5 + 24.0 + 25.0 + 23.0 + 23.5 + 24.5}{10} = 23.8 \text{ °C}$$

So, our best estimate for room temperature is 23.8 °C. How do we find the precision of this value statistically?

Standard Deviation – An estimate of the reliability of the data used to calculate the mean.

Think of this as a measurement of how spread out the data is around the mean. The formula is:

$$\sigma_x = \sqrt{\frac{1}{N-1}\sum_{i=1}^{N}(x_i - \bar{x})^2}$$

Here, we're squaring the difference between each value and the mean value, summing over all the data, dividing by $N - 1$, and then taking the square root. (Remember that taking the square root of something is equivalent to raising that same something to the one-half power.) For our data above:

$$\sigma_T = \sqrt{\frac{1}{10-1}\sum_{i=1}^{10}(T_i - \bar{T})^2}$$

$$= \left[\frac{1}{9}\{(24.0 - 23.8)^2 + (23.5 - 23.8)^2 + (24.0 - 23.8)^2 + (22.0 - 23.8)^2 \right.$$
$$+ (24.5 - 23.8)^2 + (24.0 - 23.8)^2 + (25.0 - 23.8)^2 + (23.0 - 23.8)^2$$
$$\left. + (23.5 - 23.8)^2 + (24.5 - 23.8)^2\}\right]^{\frac{1}{2}}$$

$$= \left[\frac{1}{9}\{(0.2)^2 + (-0.3)^2 + (0.2)^2 + (-1.8)^2 + (0.7)^2 + (0.2)^2 + (1.2)^2 \right.$$
$$\left. + (-0.8)^2 + (-0.3)^2 + (0.7)^2\}\right]^{\frac{1}{2}} = 0.856$$

The standard deviation is the average uncertainty *of each individual measurement*. Notice that the new uncertainty is larger (0.9) than the uncertainty from the precision (0.5) of the thermometer alone. The standard deviation is a better measurement of the uncertainty in the data because it accounts for **all** of the random error in the experiment. You use the standard deviation to report the variation in individual measurements as long as it is LARGER than the precision. If, by some chance the precision is larger than the standard deviation, you use the precision to report the variation in individual measurements because your uncertainty has to be at least as large as your precision error.

Now for something that may seem a little strange. The **mean** or average value is a better estimate of the "true" value than the individual measurements. After all, this is why we take multiple measurements of data. The mean actually has a smaller uncertainty associated with it than each individual data point.

Standard Deviation of the Mean – The uncertainty of the mean value, or the error on the mean value.

The formula for standard deviation of the mean is given by:

$$\sigma_{\bar{T}} = \frac{\sigma_T}{\sqrt{N}} = \sqrt{\frac{1}{N(N-1)} \sum_{i=1}^{N} (x_i - \bar{x})^2}$$

This is easy enough to calculate once you have the standard deviation. For the data above the standard deviation of the mean is:

$$\sigma_{\bar{T}} = \frac{0.856}{\sqrt{10}} = 0.271$$

Therefore, your best estimate for the temperature data is:

$$\bar{T} \pm \sigma_{\bar{T}}$$

$$23.8 \pm 0.3 \,°C$$

Notice that the final uncertainty is much better than the uncertainty for each data point. In fact, the final uncertainty will continue to get better as the amount of data taken increases.

Vectors

Scalars versus Vectors

In the engineering and scientific fields, numbers are always used as measurements of something. In other words, numbers stand for "real" data. Sometimes it is enough to know how big something is, or the *magnitude* of a property. For example, you might measure the height of a door, the mass of an apple, the density of steel, or the temperature of a room. These measurements all have a size (magnitude) and a unit of measurement associated with them. They are called **scalars**.

> **Scalar** – A property defined completely by a number and a unit.

Sometimes we also need to know in which direction the property is pointed. For example, you might want to know not only how fast a plane is moving (its speed, a magnitude) but also in which direction. The speed and direction together is called the velocity of an object. Force, momentum, displacement, and electric field are all quantities which are meaningless without both magnitude and direction. Properties which require both magnitude and direction are called **vectors**.

> **Vector** – A property defined completely by a magnitude (number and unit) and a direction.

Vectors are central to many engineering fields including mechanics and electrodynamics. Since vectors are not ordinary numbers; we have to redefine what we mean by addition, subtraction, multiplication, and so on just like we did for matrices.

At this level we will only talk about adding vectors, vector components, and multiplying vectors by scalars.

Graphical Addition of Vectors

Given two vectors of arbitrary units:

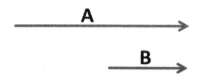

Let us create a new vector **C** which is the addition of **A** and **B**, or **C** = **A** + **B**. At first we will look at how to do vector addition graphically; later we will learn how to do this process numerically with basic trigonometry and algebra. It is important to learn the graphical method because it will help you visualize what the answer should look like. It's also important to learn the numerical method because eventually the problems will get too hard to draw out by hand. At this level however, drawing gives a qualitative method to check numerically obtained results against.

Here is the graphical process:

1. Draw vector **A** to scale.
2. Draw vector **B**, placing the tail of vector **B** at the tip of vector **A**. (The tail is the end without the arrow. The tip is the end with the arrow.)
3. The **resultant** (or vector **C**) is drawn from the tail of **A** to the tip of **B**.

So given **A** and **B** from above, **C** = **A** + **B** gives:

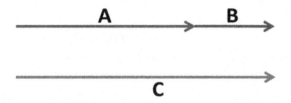

One way to think about this is that I drive 30 miles east (**A**), and then drive another 15 miles east (**B**). The resultant of this effort was that ultimately I ended up 45 miles east of where I started (**C**).

Let's look at another example.

Given

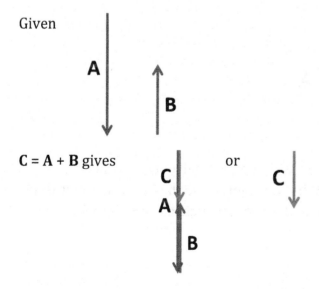

C = **A** + **B** gives   or

125

An analogy here could be that I drove 30 miles south (**A**), and then drove 20 miles north (**B**). The resultant of this effort was that ultimately I ended up 10 miles south of where I started (**C**). *Note that we didn't say, "I drove a total of 50 miles." For the purposes of vectors, we don't care what path we took to get to our destination: We just care where we ended up with regards to where we started. (The path itself is, for our purposes, irrelevant.)*

These are one-dimensional examples, intended to illustrate a couple of points. Notice in the first example the resultant is larger than the two initial vectors and in the second example the resultant is shorter than the original vectors. The resultant's length (magnitude) depends on the magnitude and direction of the original vectors. Also, where a vector begins and ends is completely arbitrary. (Whether I'm in Kansas or Peru, if I drive 30 miles south and then 20 miles north, I'm going to end up 10 miles south of where I started.) The only things defined for a vector are its magnitude and direction.

Now we will look at a two-dimensional example.

Given

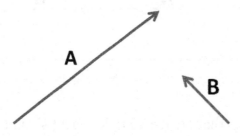

C = **A** + **B** gives

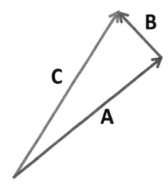

Referring back to our analogy, the only difference is that now I'm driving northeast or northwest, or some other compass direction. What is important is that it doesn't matter if the vectors are one-, two-, or even three-dimensional; the graphical addition process remains unchanged.

Be careful when adding vectors graphically that you don't connect them tail-to-tail or head-to-head. They must be connected head-to-tail.

*This does not give you **A** + **B**. Do not do this.*

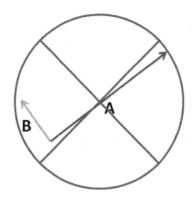

Using the graphical method and the previous A and B vectors, I can quickly show that vector addition is commutative (i.e. **A** + **B** = **B** + **A**).

C = **B** + **A** gives

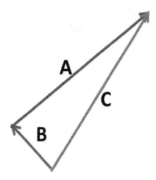

It does not matter in what order we add **A** and **B**; we get the same resultant regardless. Similarly it is relatively easy to show that vector addition is associative.

$$(\mathbf{A} + \mathbf{B}) + \mathbf{C} = \mathbf{A} + (\mathbf{B} + \mathbf{C})$$

The commutative and associative properties mean that we can add vectors together in whatever order we want and still get the same answer. This also allows us to add three, four, or more vectors together using the same graphical method.

For multiple vectors, draw the first vector anywhere; draw the second vector with its tail at the tip of the first; draw the third vector with its tail at the tip of the second; and so on.

For example, given

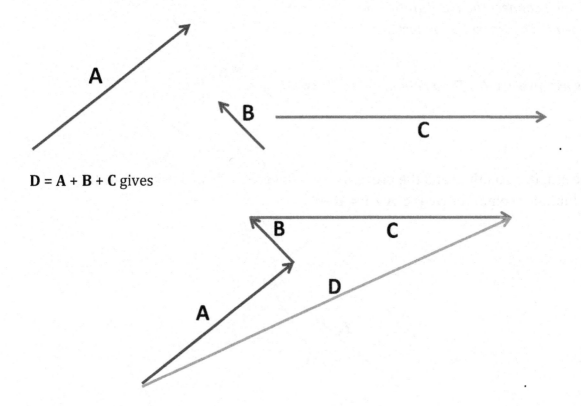

D = **A** + **B** + **C** gives

Scalar Multiplication of Vectors

Vectors can be multiplied by scalar numbers, but we have to carefully examine the possible results. Algebraically scalar multiplication looks like:

$$\mathbf{D} = c\mathbf{A}$$

The magnitude of **D** is the constant c times the magnitude of **A**. If $c > 0$, the direction of **D** is the same as the direction of **A**. If $c < 0$, the direction of **D** is rotated by 180° (or π radians) which is the opposite direction from the original direction of **A**.

Let us look at a graphical example. Consider **B**, 2**B**, and -**B** shown below.

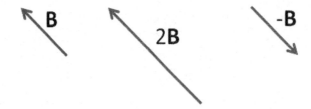

Subtraction of Vectors

Due to the properties of scalar multiplication, vector subtraction can be thought of as scalar multiplication by -1 and then vector addition. Algebraically,

$$C = A - B = A + (-1 \cdot B)$$

Graphically, given

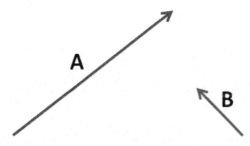

then $C = A - B$ is

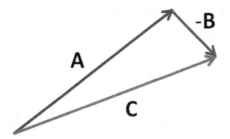

Linear Combinations of Vectors

Now, using everything we've covered so far, we can combine vectors using addition, subtraction, and scalars. For example let **A** and **B** be vectors such that

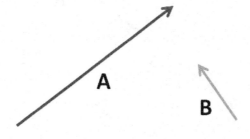

Then **C = A + 2B** is

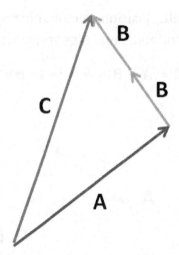

Much of what works with algebra you can expect to work with vectors. For example, if **C = A − 2B** and **D = 2B − A**, we can expect **C** and **D** to be the same size vector, just pointed in opposite directions (**C = -D**). To illustrate, we will now show that **A - 2B = -(2B - A)**. Let **A** and **B** be the vectors above, then **C = A - 2B** gives

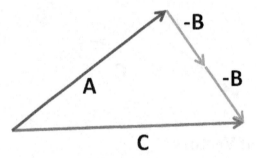

and **D = 2B - A** gives

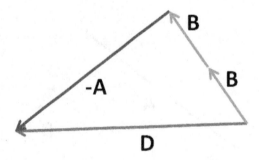

Notice that **C = -D** proving that **A - 2B = -(2B - A)**.

The graphical addition method is an excellent tool to use to visualize qualitatively the results of vector manipulation; however, we need a way to do vector math without relying on drawings, rulers, and protractors. For that, we will delve into the world of trigonometry and break the vectors into components, both of which will make manipulating vectors algebraically much easier.

Resolving a Vector into Components

To deal with vectors algebraically, we have to break them into components. Any vector can be written as the addition of two perpendicular components. So, you can make any vector the addition of an *x*-directed vector and a *y*-directed vector. These are called the *x* and *y* components of the vector. For example, $\mathbf{A} = \mathbf{A}_x + \mathbf{A}_y$

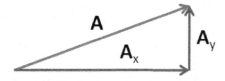

We separate vectors into components because when adding multiple vectors the *x* and *y* components can be added independently. Let $\mathbf{B} = \mathbf{B}_x + \mathbf{B}_y$

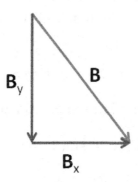

Then if **C** = **A** + **B**

$$\mathbf{C} = \mathbf{C}_x + \mathbf{C}_y$$

where

$$\mathbf{C}_x = \mathbf{A}_x + \mathbf{B}_x \qquad \mathbf{C}_y = \mathbf{A}_y + \mathbf{B}_y$$

Vectors

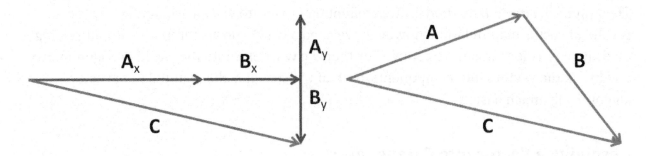

This works for adding any number of vectors together. You can calculate the components of the vector mathematically using trigonometry as long as you know the angle (direction) and length (magnitude). The process is exactly the same as getting Cartesian x and y coordinates from r and θ polar coordinates.

Given vector **A** of magnitude A and direction θ, you can get the magnitude and direction of the x and y components of **A** from

$$A_x = A \cos \theta \qquad A_y = A \sin \theta$$

As long as θ is measured counterclockwise from the positive x-axis, the directions of A_x and A_y are given by the above equations.

Let us say that we have a vector which has a magnitude of 250 N and a direction of 230°. (Another way to write that would be 250 N at 230°.) To find the x-component use $A_x = A \cos \theta$, giving

$$A_x = (250 \text{ N}) \cos(230°) \approx -160 \text{ N}.$$

(Answer rounded to 2 sig figs.) To find the y-component use $A_y = A \sin \theta$, giving

$$A_y = (250 \text{ N}) \sin(230°) \approx -19 \text{ N}.$$

Thus, we have the x and y components of the vector.

Practice finding the x and y components of the following vectors:*

a. 20.0 m at 30.0°
b. 50.0 N at 240°
c. 70.0 m/s at 135°

Vectors

Unit Vectors

To write a vector in component notation you have to use **unit vectors**. A unit vector is a vector with length (magnitude) of **one**. There is a unit vector associated with each direction: $\hat{i}$ for the positive *x*-axis, $\hat{j}$ for the positive *y*-axis, $\hat{k}$ for the positive *z*-axis

These unit vectors are pronounced: "i-hat," "j-hat," and "k-hat." So **A** from above would be written

$$\mathbf{A} = A_x\hat{i} + A_y\hat{j} = -160\hat{i} - 19\hat{j}$$

A three-dimensional vector, **B**, would be written

$$\mathbf{B} = B_x\hat{i} + B_y\hat{j} + B_z\hat{k}$$

(Getting the components for a three dimensional vector is more complicated and requires a working knowledge of spherical coordinates.)

Once you have all of your vectors written as components, adding vectors together is just a matter of adding up all of the *x*-components to get the *x*-component of the resultant and add all of the *y*-components to get the *y*-component of the resultant. So if

$$\mathbf{C} = \mathbf{C}_1 + \mathbf{C}_2 + \mathbf{C}_3 + \cdots + \mathbf{C}_N = \sum_{i=1}^{N} \mathbf{C}_i$$

then

$$C_x = \sum_{i=1}^{N} (C_i)_x \text{ and } C_y = \sum_{i=1}^{N} (C_i)_y.$$

For example if

$$\mathbf{A} = 3\hat{i} - 4\hat{j} \text{ and } \mathbf{B} = -2\hat{i} + 6\hat{j}$$

then if **C** = **A** + **B**

$$\mathbf{C} = (3\hat{i} - 4\hat{j}) + (-2\hat{i} + 6\hat{j}) = \hat{i} + 2\hat{j}.$$

Notice that you treat the i-hats and j-hats like normal variables. Try a few yourself. Add[†]

a. $2\hat{i} - 4\hat{j}$ and $3\hat{i} + 5\hat{j}$
b. $\hat{i} - 4\hat{j}$ and $3\hat{i} + 4\hat{j}$
c. $3.7\hat{i} - 4.5\hat{j}$ and $-3.1\hat{i} + 2.6\hat{j}$

Finding the Magnitude and Direction of a Vector

If we are given a vector with *x* and *y* components, sometimes we need to get back to magnitude direction notation. To achieve this, use the following formulas:

$$C = \sqrt{C_x^2 + C_y^2}$$

$$\tan \theta = \frac{C_y}{C_x}$$

You have to be careful when using the tangent function. When taking the inverse tangent on your calculator you will only get values between -90° and 90° (-π/2, π/2 radians). Unfortunately, this is only half of the unit circle. You have to deduce the correct angle using trigonometry (i.e. you need to be smarter than your calculator). There are basically four rules:

1. If the *x* component is positive, θ = calculator value
2. If the *x* component is negative, θ = calculator value + 180° (or + π)
3. If the *x* component is zero and the *y* component is positive, θ = 90° (π/2 radians)
4. If the *x* component is zero and the *y* component is negative, θ = -90° (-π/2 radians)

If you really find yourself getting confused, just look at the answer your calculator gave you and then look at your picture. If the calculator number doesn't make sense with what you're looking at, figure out why and fix it. (This is generally good advice for pretty much everything.)

Incidentally, if you have a three-dimensional vector, you can still find its magnitude (though not its direction) using a similar formula to the 2D version.

$$C = \sqrt{C_x^2 + C_y^2 + C_z^2}$$

In the previous section we added

$$\mathbf{A} = 3\hat{\imath} - 4\hat{\jmath} \text{ and } \mathbf{B} = -2\hat{\imath} + 6\hat{\jmath}$$

together to get **C = A + B**

$$\mathbf{C} = \hat{\imath} + 2\hat{\jmath}.$$

These are the *x* and *y* components of the vector **C**. If we want the magnitude and direction of **C**, we use

$$C = \sqrt{C_x^2 + C_y^2} = \sqrt{1^2 + 2^2} = \sqrt{5}.$$

So **C** has a magnitude equal to $\sqrt{5}$. To get the direction of **C** use

$$\tan\theta = \frac{C_y}{C_x} = \frac{2}{1} = 2$$

$$\theta = \tan^{-1}(2) \approx 63.435°$$

Since the *x* and *y* components of **C** are both positive, this is the correct direction for **C**.

Let's try an example given the magnitude and directions of our initial vectors. Let **A** be 27 m at 160° and **B** be 45 m at 250°. First we need to find the *x* and *y* components of **A** and **B**. So

$$A_x = (27 \text{ m})\cos(160°) \qquad A_y = (27 \text{ m})\sin(160°)$$

$$A_x \approx -25.4 \text{ m} \qquad A_y \approx 9.23 \text{ m}$$

$$\mathbf{A} = -25.4\hat{\imath} + 9.24\hat{\jmath}$$

and

$$B_x = (45 \text{ m})\cos(250°) \qquad B_y = (45 \text{ m})\sin(250°)$$

$$B_x \approx -15.4 \text{ m} \qquad B_y \approx -42.3 \text{ m}$$

$$\mathbf{B} = -15.4\hat{\imath} - 42.3\hat{\jmath}$$

Thus,

$$C_x \approx (-25.4 - 15.4)\text{m} \qquad C_y = (9.23 - 42.3)\text{m}$$

$$C_x \approx -40.8 \text{ m} \qquad C_y \approx -33.1 \text{ m}$$

$$\mathbf{C} = -40.8\hat{\imath} - 33.1\hat{\jmath}$$

Now we have the *x* and *y* components for **C**. Using those we can get the magnitude and direction for **C**. To get the magnitude for **C** use

$$C = \sqrt{C_x^2 + C_y^2} \approx \sqrt{(-40.8 \text{ m})^2 + (-33.1 \text{ m})^2} \approx 52.5 \text{ m}.$$

Vectors

To get the direction of **C** use

$$\tan\theta = \frac{C_y}{C_x} \approx \frac{-33.1 \text{ m}}{-40.8 \text{ m}} \approx 0.811$$

$$\theta = \tan^{-1}(0.811) \approx 39.1°$$

WAIT! Here the *x* component of **C** is negative, so we have to add 180° to the answer. So,

$$\theta \approx 39.1° + 180° \approx 219°.$$

Thus the magnitude of **C** is 52.5 m and the direction is 219° counterclockwise from the positive *x*-axis.

Practice adding the following vectors mathematically and finding the magnitude and direction of the resultant:[‡]

   a. $2.00\hat{\imath} - 4.00\hat{\jmath}$ and $3.00\hat{\imath} + 5.00\hat{\jmath}$
   b. $1.00\hat{\imath} - 4.00\hat{\jmath}$ and $3.00\hat{\imath} + 4.00\hat{\jmath}$
   c. 40.0 m at 20.0°, and 50.0 m at 40.0°
   d. 34.0 N at 120°, 56.0 N at -50.0°
   e. 125 m/s at 180°, 35.0 m/s at 45.0°

---

[*] a.) 17.3 m, 10.0 m   b.) -25.0 m, -43.3 m   c.) -49.5 m, 49.5 m
[†] a.) $5\hat{\imath} + \hat{\jmath}$ b.) $4\hat{\imath}$ c.) $0.6\hat{\imath} - 1.9\hat{\jmath}$
[‡] a.) 5.10 at 11.3°   b.) 4 at 0°   c.) 88.6 m at 31.1°   d.) 23.3 N at -35.3°   e.) 103 m/s at 166°

Circuits

It is hard to jump into any discussion of circuits without getting into the physics of what is going on inside a circuit element. That said, we're going to make an attempt at it. As this is an introductory textbook, we are looking to help you get an understanding of how to solve some basic problems without going into the actual details of electricity and magnetism, which requires a reasonable understanding of calculus. What we cover here will introduce some of the essential topics at a mathematical level consistent with what you have seen in this book so far.

Charge, Current, and Voltage

Any discussion of circuits must begin with definitions for charge, current, and voltage. Electric charge is a basic property of matter carried by particles which can be positive or negative and which is measured in Coulombs (C). Current is defined as the amount of charge moving past a point per second and is measured in amperes, or "amps" (A). That is, since we measure charge in coulombs (C), then 1 Amp = 1 Coulomb / second.

Voltage is the energy per unit charge of electricity, but at this level of understanding it probably makes the most sense to think about it as a pressure. Voltage, measured in Volts (V) or Joules per Coulomb (J/C), is the "pressure" that a circuit element puts on a circuit, trying to make current flow.

It may be helpful to think about the idea of current and voltage with an analogy to water.[*] Consider that you have a pipe with water (current) flowing through it. What causes that current to flow is a pressure (voltage) at the beginning of the pipe. Without the pressure (voltage), the water (current) will not flow. If the pipe is empty you can still have pressure (voltage) on the end of the pipe, but no water (current) will flow.

That is, it is possible to have voltage without current, but current cannot flow without voltage. A battery, for example can have voltage. (It's a potential for something to happen.) A wire, however, is not going to have current flowing through it if it's not hooked up to a battery or another source.

Another requirement for current to flow is a "closed circuit." Thinking again about our battery and wire, the wire has to be connected to both ends of the battery for current to flow through it. If you do not complete the circuit, no current will flow.

---

[*] This can also be considered a terrible idea because particles don't actually "flow" through a circuit the way water does through a pipe, but by the time you get to the point where this really matters, you should be sufficiently mathematically sophisticated to understand why.

Kirchhoff's Current Law

You've probably heard the saying, "What comes up must come down." There is a similar saying in circuits, "What goes in must come out."[†] Kirchhoff's Current Law (KCL) says that in a circuit, any current that enters a node has to come out of it.

What is a "node" of a circuit? A node of a circuit is any place where two circuit elements come together. For our purposes in this section, we're just going to draw rectangles as our circuit elements. These rectangles could represent batteries, resistors, capacitors, inductors, or any number of other things.

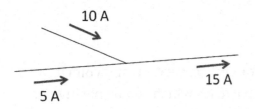

For example, consider the portion of a circuit to the left. Since we see 5 A and 10 A of current flowing into a node, the only way this could be a valid circuit would be if all 15 A of that current were flowing out, which it does.

Now consider the example to the right. We need to determine the missing value of current, I_0.

The amount of current (in amps) flowing into the node is equal to $6 + I_0$ and the amount flowing out is 12. We solve $6 + I_0 = 12$ to get $I_0 = 6$. (We are leaving the units of amperes out of the discussion for clarity, but all these measurements should eventually be labeled with A.)

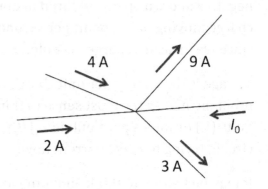

Not so bad? Let's introduce then an interesting bit that most students initially resist: when you're working through problems, it's completely okay to have a negative current! Really, it is. It just means you drew your arrows backwards. And you can leave them backwards. It's really okay. One day you're going to have a huge problem with 50 unknowns, and you don't want to have to go into every single equation and replace x's with negative x's. The easiest thing is to just leave everything the way you drew it initially and say, "This one's negative."

[†] As always there are exceptions. If you go up high enough to escape the Earth's gravity, you don't have to come back down, and eventually you can get into a situation where you'll have loss of energy through heat and other such excitement, but again, we're assuming the cow is spherical here.

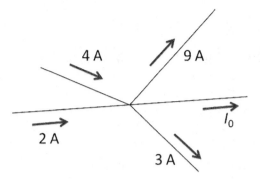

You will see an illustration of this concept to the left.

The amount of current (in amps) flowing into the node is equal to 6 and the amount flowing out is $12 + I_0$. We solve $6 = 12 + I_0$ to get $I_0 = -6$. This is the exact same circuit we had in the previous example, except that we had the arrow drawn the wrong way.

It is also possible that someone drew the arrow in backwards before you ever got there. Again, just learn to work with these kinds of issues now so that your head doesn't explode when you do get to circuits class.

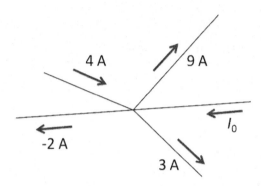

The amount of current (in amps) flowing into the node is equal to $4 + I_0$ and the amount flowing out is 10. We solve $4 + I_0 = 10$ to get $I_0 = 6$.

Again, this is the exact same circuit we had in the previous example, except that someone else drew in the arrow for the 2 A portion of the circuit backwards.

In the examples we looked at so far, there was only one node. For a typical introductory-level KCL problem, you will be presented with a circuit with multiple nodes where some currents are known and some are unknown. Consider the example below.

Here we have three unknown currents, I_1, I_2, and I_3. We also have conveniently-labeled nodes 1, 2, 3, and 4. We can choose any of these nodes to begin our analysis. Picking the "wrong" node doesn't do anything except make the problem a little harder to solve.

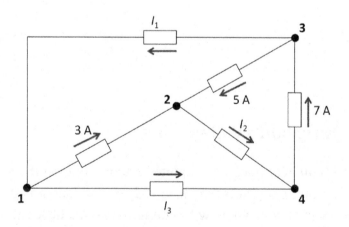

For example, if we pick node 1 to begin our analysis, we'll have I_1 coming in and $3 + I_3$ going out. We can't immediately solve $I_1 = 3 + I_3$, making our lives just a little more difficult. One alternate path for solving this problem is illustrated below.

Node 2		Node 4		Node 1	
Current In	Current Out	Current In	Current Out	Current In	Current Out
3 + 5	I_2	$I_2 + I_3$	7	I_1	3 + I_3
3 + 5 = I_2		$I_2 + I_3 = 7$		$I_1 = 3 + I_3$	
$I_2 = 8$ A		8 + I_3 = 7		$I_1 = 3 + (-1)$	
		$I_3 = -1$ A		$I_1 = 2$ A	

Note that we didn't need to use Node 3 in our initial analysis. We can, however, use it to check our answer. For Node 3, our equation would be 7 = 5 + I_1, so I_1 = 2, which is consistent with our initial answer.

Test your understanding by finding the missing values on the circuit below.*

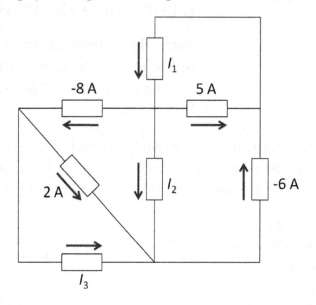

Kirchhoff's Voltage Law

If you ever become a devoted student of circuit theory, you will eventually learn about the duality of circuits which, loosely defined, means that pretty much anything you can do with current you can do with voltage, except a little different. (And if that doesn't make it crystal clear, the same principles of duality also apply to optimization and pricing schemes!)

In any case, the sister to Kirchhoff's Current Law is Kirchhoff's Voltage Law (KVL). KVL operates on the principle of the conservation of energy. Instead of working at the node, however, KVL works around a loop. The idea is that you can't go around a loop in a circuit and wind up with any more or less energy than you started with. Slightly more formally stated, the sum of voltage rises in a loop minus the sum of the voltage drops must equal zero. Any number of elements can cause a rise or fall in voltage (we'll talk about some of these in the next section), but for the moment, we will assume that these are marked for us in advance.

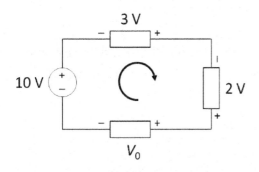

Consider the circuit to the left. There's only one loop for us to examine, and we will do so now. Let's travel through the loop clockwise, starting at the 10 V battery. We'll write an equation for the loop and say that after we go through the entire loop, we must still be at a net change of zero volts. We will use the convention of using the "first" sign we see.[‡] This means as we traverse through the loop, we will get $-10 - 3 - 2 + V_0 = 0$. We can then solve to get that $V_0 = 15$ V.

Would we get the same answer if we were counterclockwise through the loop? Absolutely! Again, starting at the battery, we would get $+10 - V_0 + 2 + 3 = 0$, which again yields $V_0 = 15$ V.

For a more complicated example, consider the circuit to the right.

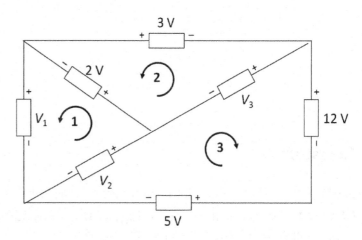

There are many loops to choose from here, just as there were many nodes in previous examples. Again, the order in which we pick the loops matters only in terms of ease of solving the problem. It won't change the actual answers. It also doesn't matter whether we go through the loops clockwise or counterclockwise.

We'll choose to solve Loop 2 first since it has the fewest number of unknowns. Again, we don't let the fact that we got a negative value for V_3 bother us. It just means that someone got the plus and minus signs switched on the drawing. All the math still works.

[‡] Technically, this means that we're "adding the drops" and "subtracting the rises," but it is all equivalent mathematically.

Loop 2	Loop 3	Loop 1
$-V_3 - 3 - 2 = 0$ $V_3 = -5$ V	$+5 - V_2 - V_3 + 12 = 0$ $+5 - V_2 - (-5) + 12 = 0$ $V_2 = 22$ V	$-V_2 + 2 + V_1 = 0$ $-22 + 2 + V_1 = 0$ $V_1 = 20$ V

If we want to check our work, we could traverse the loop that goes all around the exterior of the circuit. Starting at V_1 and going clockwise, we get -20 + 3 + 12 + 5 = 0, which is correct.

Try working the next example on your own.†

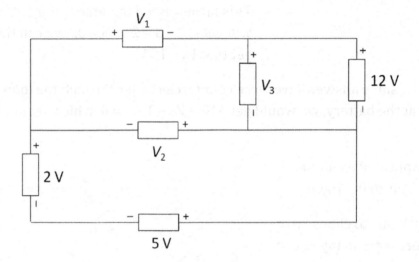

Resistors

Resistance is defined as the impedance to current flow through a circuit and is measured in Ohms, Ω. If we return to our analogy with water, adding a resistor would be like narrowing a pipe. Things that make resistance include:

- Resistors
- Wires
- Pretty much anything current flows through

That said, in an initial study of circuits, the assumption is made that there is no resistance in wires, or anything else for that matter, unless it is specifically stated as such in the problem statement.

Since it is likely that one of your first lab experiments in a Physics 2 class will be testing resistors, a good exercise is to actually learn how to "read" a resistor. A typical resistor has four bands, three of which are clustered together on one end. The bands will have different colors, and all these colors mean something. The three bands clustered together have to do with the actual resistance of the resistor, and the lone band is the tolerance of the resistor. Knowing exactly what to do with these colors may not feel straightforward at first, but it's really not all that complicated.

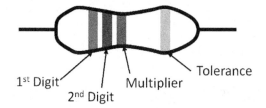

Let's determine the resistance first. The three colors should be read from left to right. Each color represents a number. The correlation between numbers and colors is given below. In this example, the first color is blue, which corresponds to a 6, the second color is violet, a 7, and the third color is red, a 2.

Our three numbers then are 6, 7, and 2. According to the diagram, we see that the first two numbers are the actual digits, and the third number is a "multiplier." The multiplier is really an exponent on the power of 10. The template for resistance, therefore, is

$$\underline{\text{Digit 1}}\ \underline{\text{Digit 2}} \times 10^{\underline{\text{Digit 3}}}.$$

As such, the resistance of the example above is $67 \times 10^2\ \Omega$, which is the same as 6700 Ω or 6.7 kΩ.

0	1	2	3	4	5	6	7	8	9
Black	Brown	Red	Orange	Yellow	Green	Blue	Violet	Grey	White

If you ever need to memorize this chart, you may find it useful to remember the mnemonic **B**ye **B**ye **R**osie **O**ff **Y**ou **G**o to **B**ecome a **V**ery **G**ood **W**ife.

Now let us look at that fourth band. The fourth band is the tolerance of the resistor, which means that the resistance may not be exactly the number provided, but should be "reasonably close." The tolerance tells how close the number should be.

Color	Tolerance
No color	± 20%
Silver	± 10%
Gold	± 5%
Red	± 2%
Brown	± 1%

In our example, the tolerance band is gold, indicating a 5% tolerance. This means that the 6700 Ω may actually measure a resistance of anything between 6365 Ω and 7035 Ω.

Practice finding the resistance and tolerance for the following color combinations. Let the last color listed represent the color of the tolerance band.‡

    a. Brown, Green, Violet, Silver
    b. Blue, Red, Red, Gold
    c. Orange, Red, Black, Red

Resistors in Series, Parallel, and Combinations

A very good exercise to help you get comfortable with dealing with resistors is to look at how they work when they are connected together. First, however, we need to establish definitions for *series* and *parallel*.

Two or more elements are said to be in *series* if any pair only shares one node between them. An easier way to think about it is, if elements are in series, then you have to go through one to get to another. Two or more elements are said to be in *parallel* if all the elements share the exact same two nodes. That is, if you think about two nodes being A and B, if they share elements in parallel, you can get from A to B multiple ways.

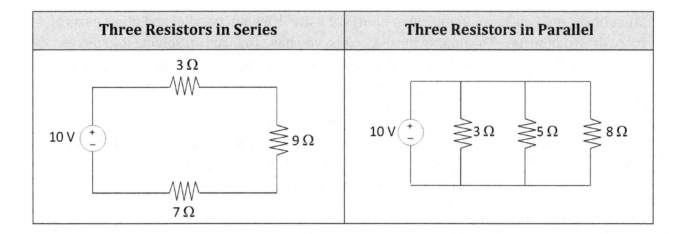

Whenever you have multiple resistors in series, it's important to recognize that the *entire* current traveling through the circuit will eventually have to go through every resistor in the circuit. That means that the current will experience a total resistance equal to the sum of all the resistors that it passes through. In the example series above, the "equivalent resistance" of the three resistors is their sum, or 19 Ω.

The equivalent resistance of *n* resistors in series is the sum of the resistances of all the resistors:

$$R_{total} = R_1 + R_2 + \cdots + R_n$$

Whenever you have multiple resistors in parallel, things get a little more complicated. Depending on the pressure (voltage) behind the current and the resistance in front of it, different proportions of the current will go through different resistors. In the example above, not all the current will just travel through the 3 Ω resistor, but more will travel through that one than through the 8 Ω resistor. The mathematical formula to describe this is not as easy as just summing them up. In fact, the total equivalent resistance is always going to be smaller than the lowest resistance. In our parallel example, the equivalent resistance is about 1.52 Ω.

The equivalent resistance of *n* resistors in parallel is given by the following equation:

$$\frac{1}{R_{total}} = \frac{1}{R_1} + \frac{1}{R_2} + \cdots + \frac{1}{R_n}$$

Another important item to note about elements in series and in parallel is the following (they are off-shoots of KCL and KVL):

- All elements in series have the same current
- All elements in parallel have the same voltage

Sometimes you will have circuits where parts of it are drawn in parallel and other parts of it are drawn in series. To simplify these circuits, you have to focus on solving it in pieces. The rule of thumb here is that if you can't figure out what you're supposed to do with the set of resistors that you're looking at, then you need to pick another set of resistors.

Consider the example below, which is essentially just our previous two examples thrown together. Let's say we decide we want to simplify the 8 Ω and the 7 Ω resistor. But after staring at it for 30 minutes, we're still stumped about how to start. That means we're not starting in the right spot. Unlike KCL and KVL, where you start matters.

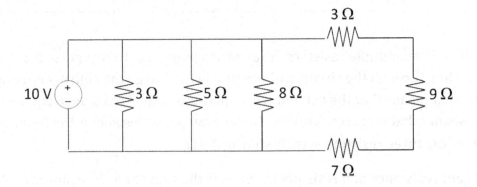

As we continue to stare at the circuit, we do see that we still have three resistors in series. We know we can replace resistors in series by one resistor represented by the sum of their resistances. We will therefore replace the 3 Ω, 9 Ω, and 7 Ω resistor with one resistor that has a resistance of 19 Ω. *(Notice that we don't include the 8 Ω resistor as part of our series. This is because the 8 Ω resistor could also be considered to be in parallel with the 5 Ω resistor. If you ever find yourself trying to argue whether or not an element is parallel or series, you're probably not starting in the right place.)*

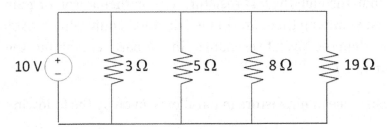

Now it is clear that all four of the remaining resistors are in parallel, so the equivalent resistance in the circuit is about 1.41 Ω.

To illustrate this point further, let's combine our two original circuits, but in a different order. In the circuit below, we now see that there are four resistors in parallel, 9 Ω, 3 Ω, 5 Ω, and 8 Ω. These simplify to one resistor with a resistance of 1.30 Ω.

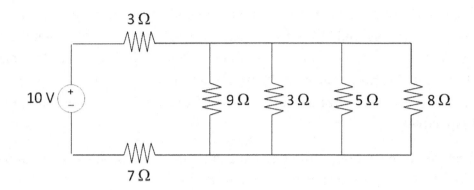

Once we have made that replacement, we now have the circuit pictured to the right. It is clear that the remaining three resistors are all in series, so we can add them up and get a total equivalent resistance of 11.3 Ω.

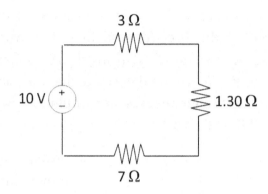

Power and Resistance

Consider again the idea of resistors as being similar to narrow pipe circumference. As the pipe narrows (as resistance increases), with the same pressure (voltage) behind it, the water (current) is going to have to flow faster to avoid a backup. If the pipe were to become wider (lower resistance), with the same pressure (voltage), the water (current) will flow more slowly. This relationship is known as Ohm's Law, and is mathematically defined as V = IR.

One application of Ohm's Law is that we can determine how much current is running through a circuit. Consider our previous example, three resistors in series.

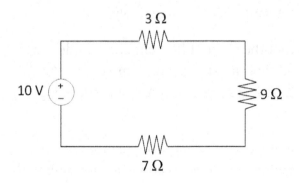

Earlier we determined that the equivalent resistance for this circuit was 19 Ω. We can use this information to figure out how much current is running through each element in the circuit. Using V = IR and substituting in V = 10 and R = 19, we can solve to get I = 0.526 A. This means that there is a current of 0.526 A running through every element in that circuit. *Remember that elements in series all have the same current.*

Another important mathematical relationship between current and voltage is $P = IV$, which says that the power (measured in Watts, W) being dissipated by a circuit element is equal to the current running through it times the voltage across it. In this example, since we know that the current running through the whole circuit is the same, we can calculate the power dissipated by each resistor as 1.58 W, 4.74 W, and 3.68 W, for the 3 Ω, 9 Ω, and 7 Ω resistors, respectively.

Working through these problems in series isn't so tough, but looking at them in parallel can be a little more tricky. Consider our previous example.

Remember that we determined that we could replace the three resistors in parallel with one 1.52 Ω resistor. Next, using the relationship that $V = IR$, we can determine that the current going through that resistor would be 6.58 A. (6.58 * 1.52 ≈ 10).

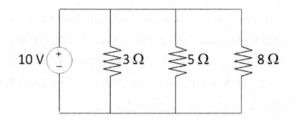

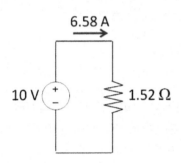

Now we can work backwards to determine the current running through each of the resistors. The voltage across the 1.52 Ω resistor is 10 V, and since we know that the voltages across any elements in parallel are equal, if we "unfold" the circuit back to its original image, then the voltage across each of the original resistors must also be 10 V.

(It wasn't strictly necessary to find an equivalent resistance to determine this, but it will help us check our answers later.)

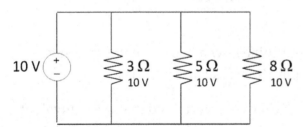

Since we know the voltage and resistance of each element, we can use $V = IR$ to find the currents. The currents running through the 3 Ω, 5 Ω, and 8 Ω resistors are 3.33 A, 2 A, and 1.25 A, respectively.

To check our results, we can use KCL. We know that the current flowing into a node must equal the current flowing out of it. We determined that the entire circuit must have 6.58 A of current, and in fact that is equal to sum of the 3.33 A, 2 A, and 1.25 A that are flowing through the individual elements.

Other questions that may relate current, power, voltage, and resistance are often more practical applications. Consider the question of asking how much wattage a light bulb with a resistance of 144 Ω gives off on a 120 V system? First, we can calculate the amount of current that the light bulb pulls using $V = IR$. Using $V = 120$ and $R = 144$, we determine that

$I = 0.833$ A. Then, since we know that $P = IV$, and using the new value of I that we just calculated, we find that $P = 100$ W. In these kinds of questions, it is just an issue of manipulating the two equations until you eventually get what you're looking for.

Mesh Analysis

A fantastic (and potentially frustrating) feature of studying circuits is that there are often multiple ways to find answers to problems. As we've mentioned before, something that can be solved by looking at current can often be solved also by looking at voltage. Sometimes one way is easier, and sometimes another way is. To whet your appetite further, we'll introduce here another method for solving the same kinds of circuits that we did in the previous section: Mesh Analysis.

Mesh analysis is a particularly powerful tool (along with its cousin, nodal analysis, which we will not discuss here) for solving larger circuits. It does not take too much imagination to come up with a circuit where using the methods from the previous section would become extremely burdensome and overwhelming. As such, mesh analysis gives us a "shortcut" around having to collapse an entire circuit and pull it apart piece by piece.

The idea behind mesh analysis is that any loop in a circuit has some kind of current going through it. Whatever that current is, it had better be some number such that the sum of all the voltage rises minus all the voltage drops must be zero (KVL). (That current could of course be zero or negative, as we've discussed.) What we can do is look at each individual loop, solve for the current running through it, and then figure out how all those individual results are related at the end.

The best way to teach this is probably through example, so let's use our tried-and-true example of the three resistors in series. We know that the voltage drop across any resistor is equal to IR. Since we don't know the current through this loop, we're going to just assign it a variable name, I_1. Traversing the loop clockwise starting at the battery, our equation will be:

$$-10 + 3I_1 + 9I_1 + 7I_1 = 0$$

Simplifying and solving this equation yields that $I_1 = 10/19$ A, or $I_1 = 0.526$ A, which is the same solution we got before. That is, the current that this loop contributes to the circuit is 0.526 A in the direction that the arrow is pointed.

For completeness, let's also look at what happens when the current arrow is drawn the wrong direction. The new equation is now:

$$+10 + 7I_1 + 9I_1 + 3I_1 = 0$$

Solving, we get that $I_1 = -0.526$ A. That is, the arrow must be pointed the opposite direction as the "true" current flow, since our current value is negative!

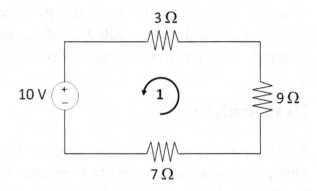

Notice that there is no variable for current next to the 10 in the equation that represents the battery charge. Why? This is because we are multiplying *current times resistance* to get the voltage drop across each resistor. The battery value is already in volts, so we just leave it as is!

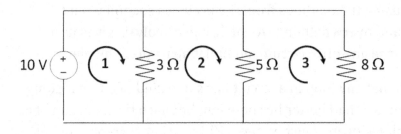

Things definitely get more interesting when we have multiple loops, and this is where the power of this method comes into play. Consider the circuit with three resistors in parallel.

Label the three currents running through loops 1, 2 and 3 as I_1, I_2, and I_3, respectively. At this point, let's consider what's actually happening here by focusing in on the 3 Ω resistor. Notice that current I_1 is running *down* through the resistor at the same time that the current I_2 is running *up* through the resistor. How can both of these be happening?

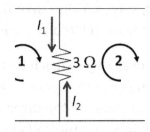

Basically the idea is that both of them *are* happening. If I_1 were 10 A (down) and I_2 were 3 A (up), then the *actual current* flowing through the 3 Ω resistor would be 7 A (down). This should remind you of vectors a little bit.

To account for this, whenever we travel through a loop, we'll take into account not just the current provided by the loop that we're currently examining, but also any currents contributed by other adjacent loops. Notice in the example below how the 8 Ω resistance is only multiplied by I_3, since I_3 is the only current traveling through that resistor.

Loop 1	Loop 2	Loop 3
$-10 + 3(I_1 - I_2) = 0$ $3I_1 - 3I_2 = 10$	$3(I_2 - I_1) + 5(I_2 - I_3) = 0$ $-3I_1 + 8I_2 - 5I_3 = 0$	$5(I_3 - I_2) + 8I_3 = 0$ $-5I_2 + 13I_3 = 0$

At this point, we have a system of equations with three variables. Using the reverse row echelon method (or any other method), we find that $I_1 = 6.58$ A, $I_2 = 3.25$ A, and $I_3 = 1.25$ A.

You may object, saying that the last time we did this exact same problem, we got that the current running through the 3 Ω resistor was 3.33 A. But remember, the current running through the 3 Ω resistor is not I_1; instead, it is actually $I_1 - I_2 = 6.58 - 3.25$, which is 3.33 A, our original answer! The current running through the 5 Ω resistor is $I_2 - I_3 = 3.25 - 1.25 = 2$ A, also consistent with our earlier answer. Finally, the current running through the 8 Ω resistor is I_3, or 1.25 A.

Now that we know the current in each element of the circuit, we can determine the power dissipated by each element, find the voltage across each element, or calculate any other number of properties as requested.

There are two ways that this can trip up a student at this level. First, remember not to multiply a voltage source by the current. (It's already a voltage – multiplying it by a current would make it a power!) Second, be sure to watch the direction of the arrows. If the current through a resistor coming from each loop is in the same direction, you will add the currents together, not subtract them as we did in this example.

To thoroughly illustrate the power of this method, let us look at a final example.

Circuits

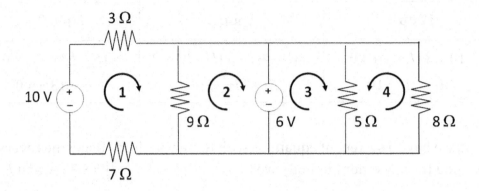

Loop 1	Loop 2	Loop 3	Loop 4
$-10 + 3I_1 + 9(I_1 - I_2) + 7I_1 = 0$	$9(I_2 - I_1) + 6 = 0$	$-6 + 5(I_3 + I_4) = 0$	$8I_4 + 5(I_3 + I_4) = 0$
$19I_1 - 9I_2 = 10$	$-9I_1 + 9I_2 = -6$	$5I_3 + 5I_4 = 6$	$5I_3 + 13I_4 = 0$

These equations yield that $I_1 = 0.4$ A, $I_2 = -0.267$ A, $I_3 = 1.95$ A, and $I_4 = -0.75$ A. We can now calculate the current, voltage and power for each of the resistors in the circuit.

Resistance (Ω)	Current (A)	Voltage (V)	Power (W)
3	(I_1) 0.4 right	1.2	0.48
7	(I_1) 0.4 left	2.8	1.12
9	($I_1 - I_2$) 0.667 down	6	4
5	($I_3 - I_4$) 1.2 down	6	7.2
8	(I_4) 0.75 down	6	4.5

Note that we indicated directions for the current flow in our table. We could have just as easily indicated these directions with arrows on our circuit. Notice also that we chose to have the current across the 8 Ω resistor as 0.75 down instead of -0.75 up. Notice also that the voltage across the 9 Ω, 5 Ω, and 8 Ω resistors are all 6 V. That should make sense, because they're all in parallel with a 6 V battery, and we know that elements in parallel all have the same voltage.

---

[*] $I_1 = -1$ A, $I_2 = 2$ A, $I_3 = -10$ A
[†] $V_1 = -15$ V, $V_2 = 3$, $V_3 = 12$
[‡] a.) 15×10^7 $\Omega \pm 10\%$, or 150 M$\Omega \pm 10\%$ b.) 62×10^2 $\Omega \pm 5\%$, or 6.2 k$\Omega \pm 5\%$ c.) 32×10^0 $\Omega \pm 2\%$, or 32$\Omega \pm 2\%$

Logic

True and False

Before you can program, you must understand the basics of logic. Much of programming is based on the idea that every statement in a program is either true (T) or false (F).

True	False
$5 = 5$	$3 = 5$
$6 < 7$	$6 > 7$
$9 < 15$	$9 > 15$
$4 \leq 4$	$6 \geq 9$
$0 \leq 4$	$-6 \geq 9$
$8 \neq 9$	$8 = 9$

In every day conversation, it is not necessarily as easy to categorize. Statements may be true, false, or neither.

True	False	Neither
Butterflies are mammals	Chickens are vegetables.	What time is it?
I like to eat potatoes.	I am ten feet tall.	Should I get a tattoo?
Humans do not breathe nitrogen.	Humans live on Mars.	Go buy some soda.
$5 > 0$	$5 < 0$	$x + 3 = 10$

Notice that statements such as "I like to eat potatoes," or "I am ten feet tall," can be considered true or false based on the identity of the speaker. It is also possible to have true and false statements with the word "not" in them. The equation $x + 3 = 10$ is neither true nor false because x has not been defined.

And

In logic, AND is the conjunction or the intersection. In the example below, if Group A contains the people who like puppies and Group B contains the people who like kitties, then A AND B represents the subset of people who like puppies and kitties.

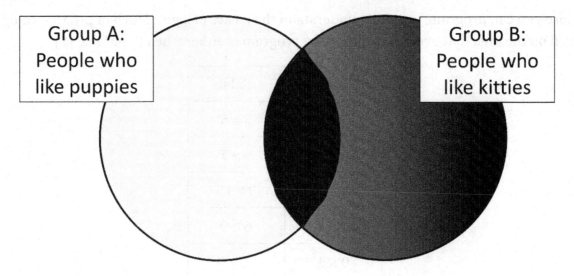

Other symbols for AND include ∧ ∩ & &&.

This relates to programming since decisions are often made based on whether or not multiple conditions exist. For example, suppose an online store was having a sale where if someone bought a USB Drive and bought a case of Powerade, they would get a 10% discount. Corresponding code might look something like this:

```
If (Bought_A_USB_Drive AND Bought_A_Case_Of_Powerade) Then
    Message   "Congratulations!  You get a 10% discount!"
End if
```

In logic, it is important to understand how statements joined with AND operate. In the example above, the customer would only get a discount if both conditions were met. For the AND statement to be true, both statements within it must be true.

For example, let p be the statement "We live on Earth." Let q be the statement "We breathe oxygen." These statements are both true. Therefore, the statement "We live on Earth AND we breathe oxygen," is also true. That is, p AND q is True, or $p \wedge q = T$.

Continuing this illustration, let p be the statement "We live on Earth." Let q be the statement "We breathe nitrogen." Only one of these statements is true, so the statement "We live on Earth AND we breathe nitrogen," is false. That is, p AND q is False, or $p \wedge q = F$.

Finally, let p be the statement "We live on Mars." Let q be the statement "We breathe nitrogen." Neither of these statements is true, so the statement "We live on Mars AND we breathe nitrogen," is still false. (Two wrongs don't make a right.) That is, p AND q is False, or $p \wedge q = F$.

We can summarize this information in a truth table.

p	q	$p \wedge q$
T	T	T
T	F	F
F	T	F
F	F	F

Or

Another important logical operator is OR, is the disjunction or the union. In the example below, if Group A contains the people who like puppies and Group B contains the people who like kitties, then A OR B represents the subset of people who like puppies or kitties (or both).

Other symbols for OR include ∨ ∪ | ||. (These last symbols are on the backslash key above the enter key on most keyboards.)

Continuing the examples from above, suppose an online store was having a sale where if someone bought a USB Drive or spent over $100, they would get a 10% discount. Corresponding code might look something like this:

```
If (Bought_A_USB_Drive OR Spent_Over_100_Dollars) Then
     Message   "Congratulations! You get a 10% discount!"
End if
```

Statements joined with OR operate differently than those with AND. In the example above, the customer can get a discount if either condition is met. For the OR statement to be true, only one of the statements within it has to be true.

For example, let p be the statement "We live on Earth." Let q be the statement "We breathe oxygen." These statements are both true. Therefore, the statement "We live on Earth OR we breathe oxygen," is also true. That is, p OR q is True, or $p \vee q = T$.

Continuing this illustration, let p be the statement "We live on Earth." Let q be the statement "We breathe nitrogen." Only one of these statements is true, but that's good enough, so the statement "We live on Earth OR we breathe nitrogen," is true. That is, p OR q is True, or $p \vee q = T$.

Finally, let p be the statement "We live on Mars." Let q be the statement "We breathe nitrogen." Neither of these statements is true, so the statement "We live on Mars OR we breathe nitrogen," is still false. (We have to tell the truth at least once.) Therefore, p OR q is False, or $p \vee q = F$.

We can also summarize this information in a truth table.

p	q	$p \vee q$
T	T	T
T	F	T
F	T	T
F	F	F

Logic

Keeping the Symbols Straight

You take what you can get when it comes to mnemonic devices. As such, with the authors' apologies, you may find these helpful.

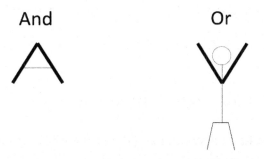

And

∧

"A" for And...

Or

∨

This character shrugs his shoulders – he doesn't care, one way "or" the other is fine!

Negatives

Writing the negative of a statement is basically writing its opposite. We usually do this by adding or removing the word, "Not." When writing the negative of statements, you make a true statement false, or a false statement true. Remember, the point is not to "correct" the statement – merely to write the opposite of it.

Statement	Negative of Statement
Today is Tuesday.	Today is not Tuesday.
It is hot outside.	It is not hot outside.
I ate hot dogs and hamburgers.	I did not eat hot dogs and hamburgers.
I do not like bugs.	I do like bugs.
$6 > 9$	$6 \leq 9$
$-6 \geq 9$	$-6 < 9$
$8 = 9$	$8 \neq 9$
$4 < 4$	$4 \geq 4$
$0 \leq 4$	$0 > 4$
$9 \neq 9$	$9 = 9$

Notice that the negative of statements like "It is hot outside," is not "It is cold outside." This is again because we are not correcting the statement, we are simply writing the opposite of it.

The sign for negatives in logic can vary between programming languages. Symbols for NOT include ~ ! <>. For the purposes of this discussion, this chapter will use the ~ symbol.

It is important to remember that ~F = T and ~T = F.

Precedence of Operators and Logic Statements

Just as mathematical operators have precedence (Parentheses, Exponents, Multiplication/Division, Addition/Subtraction), logical operators also have precedence. The precedence of operators is as follows:

1. Negation
2. And
3. Or

After precedence is followed, statements are analyzed from left to right. Armed with this knowledge one can now simplify logical statements. For example,

F ∧ T ∨ **~F**	*Simplify the negative operator first.*
F ∧ T ∨ T	*Next, simplify the AND.*
(F ∧ T) ∨ T	*Use parenthesis to emphasize precedence.*
F ∨ T	*F and T is F.*
T	*F or T is F.*

Additional examples follow, with the section to be simplified emphasized in **bold** and its simplification in ***bold italics***:

- **F ∧ T** ∨ F ∨ T
  F ∧ T ∨ F ∨ T = ***F*** ∨ F ∨ T
  F ∨ F ∨ T = ***F*** ∨ T
  F ∨ T = ***T***
- T ∧ **~T** ∨ **~T**
  T ∧ **~T** ∨ **~T** = T ∧ ***F*** ∨ ***F***
  T ∧ F ∨ F = ***F*** ∨ F
  F ∨ F = ***F***

Logic

- $T \land \sim F \lor T \land F$
  $T \land \sim F \lor T \land F = T \land T \lor T \land F$
  $T \land T \lor T \land F = T \lor T \land F$
  $T \lor T \land F = T \lor F$
  $T \lor F = F$

Essential Writing Skills

Engineering Design Documents

Introduction

Only a very few lucky engineers will get to participate in a design process from the moment of conception to the retiring of a product. This may depend on the industry. For example, in telecommunications, a product life cycle may only be 18 months, while in other industries, like defense, a product life cycle may take 30 years. Even for those industries with shorter concept-to-market times, however, there are usually teams of engineers who focus on different aspects of development, including initial design, development, marketing maintenance, upgrades, and the eventual phasing out of a product or model. Complicated products may have multiple groups working on different aspects of design and support, such as safety and human factors, logistics, reliability, or customer support.

Different companies will also have different methods of documenting the design process, each with its own requirements and expectations. In some companies, engineers must write huge proposals and win contracts before the company will officially endorse a design, while others end up documenting "after the fact" if at all.

The best way to be prepared for your career is to get exposure to as many aspects of the design process as possible. In this text, we will introduce one method of going through the design and documentation process. There are several purposes behind this kind of documentation:

Technical archive. In ten years, if your company decides to recreate this product, the documentation that was completed during the design process will help your company in following your methodology and in avoiding any mistakes or pitfalls your team experienced during the initial design process.

Patent defense. When filing or defending a patent, the more documentation you have of your design process the better. Note that the kind of documentation described here would not provide the kind of concrete evidence required in patent litigation, but practicing creating these kind of reports will get your mind in the right gear.

Improved design process. Mature companies will often analyze the design process itself to determine what methods work and what methods produce faulty or otherwise bad products. By documenting your thought process during design, you become a more precise engineer and designer.

Basic Layout

Most design documents will follow this kind of layout:

1. Executive Summary
2. Problem Statement
3. Research
4. Operational Need and Requirements
5. Concept Exploration
6. Analysis
7. Recommendations
8. Future Opportunities

In more complicated design processes, each one of these sections may be individual documents containing hundreds of pages. The rule is that the documentation should be "as long as it needs to be." Never ask your boss (or your professor for that matter!), "How many pages should this be?" Write the document with exactly as much detail as is necessary to describe your process, no more, no less. (The exception is for government proposals, which have very specific page requirements.)

We will explore each of these sections further. As you write each section of the document, you must be as detailed as possible. Rather than saying, "We explored many options," list those options. Instead of saying, "The average customer has three children," create a table or graph where you summarize the data that led to that conclusion.

You will probably write these sections starting with 2 and moving to 8. Do not expect this to be an "I wrote it once so I'm done" kind of experience. You will very likely revisit and revise sections, particularly going through sections 3, 4, and 5. Design is an iterative process.

The executive summary should be the last thing you write, since you cannot know your design recommendation until you have gone through the process.

Executive Summary

Audience: General.

The executive summary should be about 150 – 300 words and should succinctly summarize the problem to be solved, the research methodology, significant findings, the final solution, and any open issues to be resolved. The executive summary should provide enough information so that someone coming upon the report could quickly determine whether or not the information he or she may be looking for is contained within the report. An executive summary may also be provided to a potential investor.

Problem Statement

Audience: General.

The problem statement should detail exactly why the design process is being initiated. Any design can be traced at some point back to someone trying to solve a problem. (If there were no problems, no one would need to fix them!) If there is an experience that explicitly initiated the design process, such as a customer complaint or new customer expectations, that should be included in this section.

Rather than saying, "I want to design a pen that works in space," for example, you should begin the design process with, "I need to write things down while I'm in my space capsule." This distinction is very critical. If you begin the design process with a final product already in mind, you very well may end up with a solution that is less than optimal. For example, if all you need to do is write in space, a pencil would work just fine, and designing a pen is a wasted effort!

Research

Audience: Technical.

This section is exactly what it sounds like. This is where you document everything that there is to know about your problem. Important questions to answer in this section are:

- What existing market solutions are available?
- How much do these solutions cost?
- How do these solutions meet customer requirements?
- How are potential customers modifying these existing solutions to meet their needs?

Ideally in this section you will want to include details on price, dimensions, and features, as well as including images of these solutions.

Operational Need and Requirements

Audience: Technical.

This is the section where things start to get specific. You may revisit this section many times throughout the design process, particularly as you progress through the previous section, which is Research. As you discover new things, you may need to add or subtract requirements. As you add or subtract requirements, you may need to conduct more research. Write this section in two parts.

Part 1: Required Features

What features must the solution have to meet the absolute basic needs of the customer? Here is where you may list requirements such as, "Must not exceed $100 per unit" or "Must not pull more than 60W of power." If you are designing a shopping cart, you should list the maximum weight of the groceries expected in the cart. If you are designing a laptop, you should list the processing requirements.

In addition to listing these requirements, you should also consider listing the justification for them. Why do you need to meet these specific requirements?

Anything that your product is expected to do or be should be listed in this section. Later you will use this checklist to eliminate any potential designs that do not meet every single one of these requirements.

Part 2: Optional Features

This is where you talk about the things your customer would like to have, but that isn't strictly required. Potential solutions may or may not meet all of these requests, but those that meet more of them will be more likely to be chosen as the final design. Optional features could include things such as, "Should have Bluetooth capability," or "Should be customizable with favorite NFL team logo."

"Must" versus "should." In general, if you find yourself saying, "This design must blah blah," then that is a requirement that belongs in Part 1. If you say, "This design should have blah blah," then that is an option that belongs in Part 2.

Concept Exploration

Audience: Technical.

This section is where you list any potential solution ideas. You should include drawings, specifications, and utility. Depending on how detailed your design process, you may also include estimates on maintenance and support, cost, materials, and other details. List concepts even that you are not sure will be successful. You will eliminate them later in the **Analysis** section.

Analysis

Audience: Technical.

This section is strongly tied to the **Operational Needs and Requirements** section and should be completed in two parts.

Part 1: Required Feature Analysis

In the **Operational Needs and Requirements** section, you listed many requirements that were necessary for any final design. Your first task is to compare those requirements to the potential solutions you developed in the **Concept Exploration** section.

Remind your reader what the requirements were, and then create a chart to compare each of the potential solutions you developed previously with each of the requirements. Any potential solution that does not meet *all* requirements must be discarded and not used in future analysis. (Do not eliminate the solution from your entire document – this is part of the analysis process, and documenting even failed solutions is acceptable and necessary.)

Part 2: Optional Feature Analysis

Here you will compare your remaining solutions to the optional features list you developed in the **Operational Needs and Requirements** section. Your first step is to establish a "weight" or "importance" score for each optional feature. (This is similar to how a professor may weight homework at 40%, tests at 30% and the final at 20%.) You may need to have further discussions with your customer to establish the relative importance of each optional feature.

The next step will be to create "scoring functions" for each of the optional features. How will you decide whether or not each concept satisfies each optional feature? You may choose to use customer surveys, observations, analysis, or simple yes/no scores. The kind of optional feature you are scoring will directly impact how you decide to score it.

Using the scoring functions you've now created, score each solution on each optional feature. Provide any details on the methodology for computing the scores. Then use a weighted average to find a total score for that solution.

Recommendation

Audience: General.

Based on your scores, make the recommendation for the final design. Remember that a design choice should not be based on "feelings" or "opinions." If you ever find yourself writing, "We feel...," "We believe...," or similar statements, delete them immediately. Your recommendation should be based strictly on the score. Summarize the features of the design and point out technical specifications where this design scored particularly well against its competitors.

Future Opportunities

Audience: General.

This is your opportunity to describe any additional features you may not have been able to include in your design due to impediments like cost or unavailability of technology. Here you should list areas for future research or potential upgrades for the product.

Engineering Design Document Example

What follows is a good example of a simple engineering design document. This is only intended to provide student design teams inspiration for creating documents of their own and is by no means intended to be a "Perfect Example." Key features of this document are scaled drawings, specific requirements and scoring functions, and a good research section. Other features of this example document are consistent use of headings and table formatting.

Play, Snooze, & Snack

Design and Documentation by

Student A

Student B

Student C

Date

Executive Summary

A local grocery store owner came to our design team looking for a more attractive and useful shopping cart to provide to her customers who are making shopping trips with small children. We discussed the problem with the client and surveyed a select group of her customers to further define the problem. We then conducted research to determine the shopping carts used by competing grocery chains in the client's area. We surveyed customers' shopping habits and how they interacted with their carts while shopping with children. We also investigated industry standards for shopping carts. Continuing to consult with the client, we established several key parameters that any shopping cart design must meet and then listed additional features that were preferred in the design. We created five unique designs, which were then compared against the key parameters list. All but one of the initial designs met all key parameters. Further analysis, including a customer survey, allowed the team to settle upon a final design, the "Play, Snooze, & Snack," which features seating and sleeping accommodation for two small children (under 40 pounds), along with "steering wheels," three cup holders, a purse compartment, and a "Delicate Items" tray for keeping bread, eggs, and other easily-damaged products separate from the rest of the cart. We conclude this report with our findings and opportunities for future development.

Problem Statement

A client owns a chain of grocery stores. The client has expressed an interest in upgrading the shopping carts in her store so that she can compete for and attract more high-end customers. The client's stores emphasize organic foods and environmentally-friendly products and she seeks to offer a peaceful and enjoyable shopping experience.

The client has a major problem with the shopping carts that are currently available on the market. The majority of her customers (65%) are women or men with children. About 40% of these customers have a child under the age of three that accompanies them while shopping. With the child in the front seat on traditional carts, the customers are left with no room for their purses and cloth shopping bags. Also, with the front seat occupied, there is no protected place in the cart for delicate fruits, vegetables, and bread.

Other customer complaints include:

a. No place on the cart for a drink for adult
b. No place on the cart for a drink for child or children
c. No educational stimuli for child or children
d. No place to keep a grocery list and cross items off as they are retrieved (child will grab the grocery list and eat it)

Our charge is to research the situation and design a new shopping cart that meets our client's expectations and makes her customers' shopping experiences more enjoyable.

Research

Investigation of Previous Solutions

Our team investigated several shopping carts at grocery stores in the same market area (Waco, Texas). There are two competing grocery chains in the area.

Table 1: Summary of Existing Solutions

	HEB	Wal-Mart
Child capacity	2	1
Main cargo volume	9,150 cu in	10,450 cu in
Bottom tray dimensions	6,110 cu in	6,180 cu in
Other features	This design features a simple tray at the back of the cart for storing delicate items.	This is a "standard" shopping cart and has very few features.

Investigation of Cart Usage: Large Items

Large items typically include large bags of dog food, paper towels, toilet paper, and 24-packs of soda or water. Our observations showed that 90% of customers purchased a maximum of two of these items per trip. Sample dimensions of these items are given in the table below.

Table 2: Summary of Common Large Items Purchased

Item Type	Dimensions (length x width x height) inches
Dog Food (44.1 lb)	33" x 5" x 17"
Paper Towels (8 pack)	9" x 11" x 18"
Toilet Paper (24 pack)	8.5" x 9" x 22.5"
24-pack Soda (3 x 4 x 2 cube orientation)	7.5" x 10.5" x 10"
24-pack Soda (6 x 4 x 1 flat orientation)	15.5" x 5" x 10.5"

Observations of the Child-Accompanied Shopping Experience

With the client's permission, we sent several secret shoppers to her store to observe her customers' experiences. We watched a total of 5 shopping families as they progressed through the store. For consistency, we only watched shopping families with one adult and no children over the age of 3.

Table 3: Observation of Shopping Experiences

	# of Children	Observation
Family 1	1	Child sat calmly in the provided seat until child saw a rack of gummy bears, at which point child started a tantrum. Mother mollified child by opening a box of gummy bears, which she later purchased.
Family 2	2	Larger child sat in seat while smaller child was held on hip. Later during the shopping process, the smaller child was put in the seat and the larger child asked to walk. The larger child insisted on pushing the cart and inadvertently ran into a display. Parent convinced child to hold onto the cart from the side.
Family 3	1	Child sat in provided seat and insisted on holding the shopping list. Each time parent selected an item, child wanted to hold it, discarding the previously-held item by throwing it into the main cargo area of the cart, at one point damaging a loaf of bread.
Family 4	1	Child sat peacefully in provided seat, but then fell asleep. Since no lounge was provided, parent had to carry sleeping child and push cart one-handed.
Family 5	1	Child and parent talked while child sat in provided seat. Child ate an apple provided by parent and the shopping experience continued without incident until the checkout process, at which point the parent had to go to the opposite end of the cart to load the groceries onto the checkout lane. Child became anxious, no longer seeing the parent, and cried until the parent returned to the front of the cart.

Industry Standards

Grocery carts come in a variety of shapes and sizes. Typical measurements[*] and costs[†] for the "traditionally styled" shopping cart are summarized in the table below. Nesting is defined as the distance between the bottom of the basket and the shelf below.

Table 4: **Shopping Cart Dimensions**

	Small	**Medium**	**Large**
Width	19.00 in	21.65 in	22.75 in
Length	33.25 in	39.76 in	39.38 in
Height	36.00 in	40.79 in	42.00 in
Nesting	8.00 in	10.24 in	10.50 in
Basket capacity	6,200 cu in	8,847 cu in	10,300 cu in
Shopping capacity	8,600 cu in	12,385 cu in	16,480 cu in
Weight capacity	300 lbs	300 lbs	350 lbs
Price (Plastic)	$117	$152	$145
Price (Metal)	$104	$123	$135

Additional Observations

- A trip to a local women's shop revealed that the average woman's purse is about 8 x 12 x 6 inches.
- A trip to a car dealership revealed that the average cup holder is 3 inches in diameter.
- Comfort experiments with the design team and compliant family members concluded that most people are comfortable with a 15° recline for snoozing.
- Reusable grocery bags sold at the client's grocery store measure 13.6 inches wide, 6.5 inches deep, and 15.6 inches tall.
- The typical shopper at the client's grocery store purchased 10 bags of groceries.

---

[*] Source: http://www.rwrogerscompany.com/new-shopping-carts
[†] Source: http://premiercarts.com/

Operational Need and Requirements

Putting together a solution for this client is a top priority for our company. Design criteria are listed below

Part 1: Required Features

The final design must feature all of the following:

- A child-carrying capacity of at least two children at a maximum of 40 pounds each
- Storage for a purse of average size, which is 8 x 12 x 6 inches
- Storage for 10 cloth grocery bags
- Capacity to hold 10 bags of groceries, plus two large items
- Space for storing delicate or breakable items such as bread or eggs
- A cost of no more than $200 per cart

Part 2: Optional Features

Other design considerations should include one or more of the following:

- Customer appeal
- A circular drink holder 3 inches in diameter
- A flat surface inaccessible to the child seating place for writing
- Storage for a pencil or pen
- Entertainment options for children
- Sleeping options for children

Concept Exploration

We came up with five different designs. In order to meet capacity requirements, each design stays within the footprint of existing large carts and includes the standard bottom shelf. Therefore, all are based on the standard metal-frame shopping cart. All cart designs include our "Ultra-Flat Surface" on the bottom shelf, which allows items to slide smoothly and without damage to paper or plastic wrappings. Our designs also include a strong plastic clip to help keep children secure within their seats that is guaranteed for 10 years. In the drawings below, each square represents two inches.

Concept 1: The Eco-Shopper. This model features six drawers that pull out on each side of the cart. Each drawer will fit one cloth shopping bag, so the customer can fill up his or her bags while progressing through the store. There is a basket above these drawers for smaller or lighter items. The child seat in front is retractable/foldable as in standard shopping carts and accommodates two children at a maximum of 40 pounds each. This design also has a console on the handlebar which includes three cup holders and a slanted writing surface for keeping track of grocery lists. For entertainment value, there are also two "steering wheels" for each child. This model also includes a purse compartment and a special area for milk and juice.

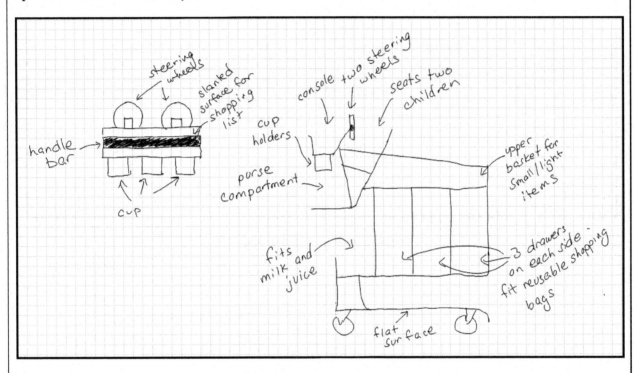

Figure 1: Concept 1 - The Eco-Shopper

Concept 2: The Co-Pilots. This model also features seating for two older children up to 60 pounds, and four chains help keep the children in. A console near the handlebar includes a cup holder and small cubby for a shopping list, phone, or coupons. There is also an upper basket for delicate items, and there is a purse compartment.

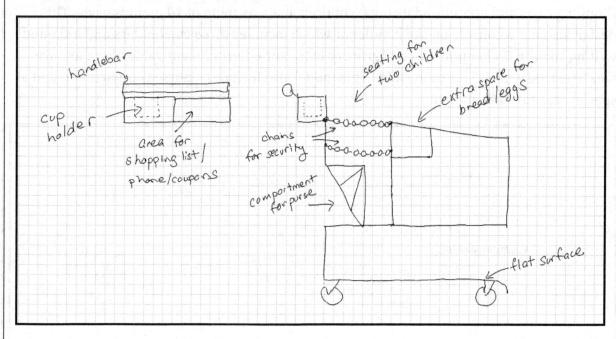

Figure 2: Concept 2 - The Co-Pilots

Concept 3: Snooze & Shop. This model features seating for three children, with two in the front seat and one older child out front. The seat for the older child includes a cup holder. The seats for the smaller children feature an additional surface that can be swiveled up to give a sleeping child something to lean against. There is also a purse compartment.

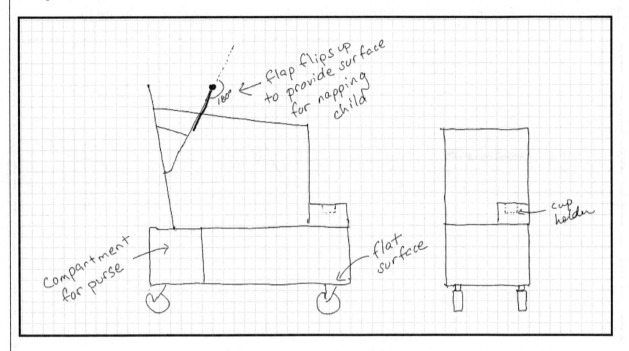

Figure 3: Concept 3 - Snooze & Shop

Concept 4: Play, Snooze, & Snack. This design features seating for two children at a maximum of 40 pounds each. The child seat in front is retractable/foldable and includes an additional surface that can be swiveled up to give a sleeping child something to lean against. There are also two "steering wheels" for each child. The console on the handlebar includes three cup holders and a slanted writing surface for keeping track of grocery lists. This model also includes a purse compartment and an upper basket for delicate items.

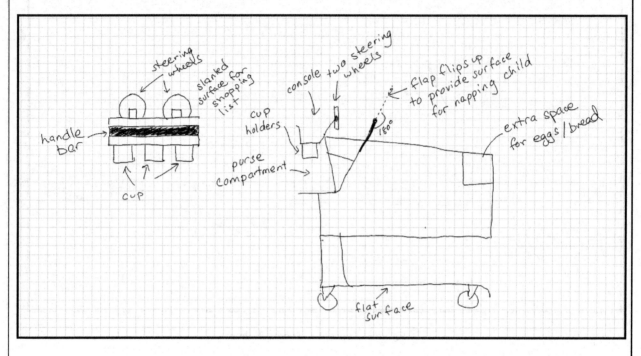

Figure 4: Concept 4 - Play, Snooze, & Snack

Concept 5: Got Your Back. This model features two child seats for larger children (up to 60 pounds each) facing opposite directions. This model may appeal to the parent who has two children who may not like to "share" space. There is also an upper basket for delicate items. The bottom section is subdivided to allow for more specific cart-loading. This design also includes a console near the handlebar with a cup holder and small cubby for a shopping list, phone, or coupons.

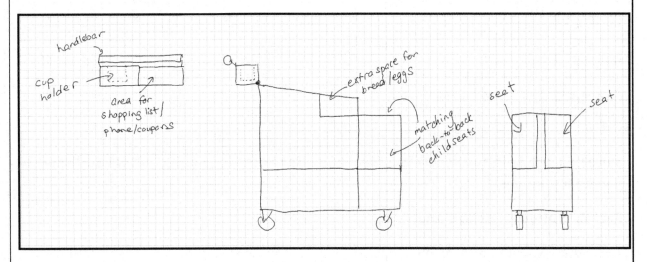

Figure 5: Concept 5 - Got Your Back

Analysis

In order to determine which design would best meet our client's needs, we performed a detailed analysis, looking at both the required features and optional features, as defined in the **Operational Needs and Requirements** section.

Part 1: Required Feature Analysis

There were five primary criteria that all designs needed to meet. Any design that did not meet all five criteria was discarded. The required criteria were:

- **"Child"** – A child-carrying capacity of at least two children at a maximum of 40 pounds each.
- **"Purse"** – Storage for a purse of average size, which is 8 x 12 x 6 inches.
- **"Bags"** – Storage for 10 cloth grocery bags. Concept 1 actually uses the cloth bags in the design, and each cart has a flat surface that can be used to store empty bags.
- **"Capacity"** – Capacity to hold 10 bags of groceries, plus two large items.
- **"Delicate"** – Space for storing delicate or breakable items such as bread or eggs.
- **"Cost"** – A cost of no more than $200 per cart.

Table 5: Critical Design Requirements						
	Child	**Purse**	**Bags**	**Capacity**	**Delicate**	**Cost**
Concept 1	✓	✓	✓	✓	✓	✓
Concept 2	✓	✓	✓	✓	✓	✓
Concept 3	✓	✓	✓	✓	✗	✓
Concept 4	✓	✓	✓	✓	✓	✓
Concept 5	✓	✓	✓	✓	✓	✓

Since Concept 3 did not meet all the design requirements, we removed it from the list of potential solutions.

Part 2: Optional Feature Analysis

To further evaluate the remaining options, we creating scoring functions based on the additional features and design considerations. Working with our client, we assigned an importance weight (percentage) to each feature and then determined a scoring function.

- **Customer appeal ("Appeal") – 50%.** This particular item is difficult to evaluate. The team noticed that there are often child-friendly carts in grocery stores, but they are not always used because shoppers will shy away from them because they "look weird" or "seem bulky." We realize that an initial reaction is critical to whether or not a customer would consider using a particular design. Once the customer uses a particular model, then they are more likely to continue using it if it meets their needs. Also, the primary motivation for our client in creating this design is to attract more customers to her store. As such, we evaluated this by presenting drawings of the four remaining concepts to 100 customers at the client's store and asked, "Which one of these carts would you be most likely to use?" We compiled a summary of the responses and divided by the maximum number of responses over all four concepts to normalize the data. Then we multiplied that value by 50 to scale the scores from 0 to 50.
- **A circular drink holder 3 inches in diameter ("Drink") – 10%.** This is easy to determine directly from the drawing. A design with at least one drink holder would earn 6 points, and earn 2 points for each additional cup holder, for a maximum of 10 points.

- **A flat surface inaccessible to the child seating place for writing ("Writing") – 5%.** This is something determined directly from the drawing. We scored this item as a "Yes" for 5 points or a "No" for 0 points.
- **Storage for a pencil or pen ("Storage") – 5%.** This is something determined directly from the drawing. We scored this item as a "Yes" for 5 points or a "No" for 0 points.
- **Entertainment options for children ("Entertainment") – 15%.** This is something determined directly from the drawing. We scored this item as a "Yes" for 15 points or a "No" for 0 points.
- **Sleeping options for children ("Nap") – 15%.** This is something determined directly from the drawing. Since some of our designs included some sleeping options and some non-sleeping options, we gave 10 points to a design which included a sleeping option, and awarded 5 additional points to designs which could accommodate more than one sleeper.

Table 6: Customer Survey Results

	Concept 1	Concept 2	Concept 4	Concept 5
Primary Choice Selection	33	13	36	18
Normalized Value	.92	.36	1.00	.50
Score	46	18	50	25

Table 7: Concept Scoring

	Max Score	Concept 1	Concept 2	Concept 4	Concept 5
Appeal	50	46	18	50	25
Drink	10	10	6	10	6
Writing	5	5	0	5	0
Storage	5	0	5	0	5
Entertainment	15	15	0	15	0
Nap	15	0	0	15	0
Total	100	76	29	95	36

Recommendation

Based on our analysis, our team recommends Concept 4: Play, Snooze, & Snack to the client. This particular design meets all the essential requirements, is most appealing to the customers in our survey, and provides the most features.

Future Opportunities

As this design matures, we may consider adding more entertainment options to the cart. The wheels may be designs to be removed and replaced with other types of toys which may be more age-appropriate. Since the design that incorporated the drawers and reusable bags was also quite appealing to customers, we suggest that future iterations of this cart consider using a similar design. As more customers move to these reusable bags, designs like this may become more and more appealing.

Grammar and Spelling Reminders

Why are we talking about this in an engineering class?

Maybe misspelling words wasn't so much of a big deal in high school. Guess what? Now, when you don't spell something right, it makes you look like you lack a basic education, which is definitely something you don't want to see in an engineer. The idea is if you can't even spell rocket, how can anyone expect you to design one?

The same thing goes for grammar. You may be called upon one day to write a proposal to win a government contract or to create a sales pitch. No one will be impressed if you have a point that states, "Its clear that our solution meets you're needs while insuring that the principle attributes are uneffected." (If you don't see five mistakes in that sentence, definitely don't skip this chapter!)

That said, clearly none of us lack a basic education – we wouldn't be engineers (or aspiring ones) if we did. Occasionally, however, some things can slip through the cracks. Correct grammar and spelling is really just an issue of memorization. Make it a point to get everything in this chapter straight. Learn it once really well, and you'll have it forever. If you'd like some more direction, visit http://www.lessontutor.com/ltgram9home.html.

Commonly Confused Words

advice	an opinion or suggestion (this one's the noun)
advise	to give an opinion or suggestion (this one's the verb)

affect	to act on; produce an effect or change in (this one's the verb)
effect	result; consequence (this one's the noun)

angel	an attendant or benevolent spirit
angle	a measurement of the relationship between two lines intersecting at a point

ensure	to secure or guarantee, or to make sure or certain o I will *ensure* my house doesn't fall down by using the proper structural support.
insure	to guarantee against loss or harm, like taking out an insurance policy in case of disaster o I will *insure* my house against fire by purchasing a policy.

farther	at a greater distance o The tree is *farther* away than the boat.

further	at or to a greater degree, as in time or as in an intensity of something
	
	    o Let's not discuss this *further*.
	    o You are in school to *further* your education.

it's	it is o *It's* really fun to study on a Friday night!
its	belonging to it o The dog wagged *its* tail.

lose	to misplace or no longer possess o I always *lose* my keys.
loose	not bound together o This connection is *loose*.

principal	something that is the highest in value or rank (or a person who is the highest in rank) <ul><li>I will serve as the *principal* investigator.</li><li>The *principal* of the school issued a decree.</li></ul>
principle	a fundamental rule <ul><li>Engineering requires a mastery of the *principles* of physics.</li></ul>

their	belonging to them
there	a location
they're	they are

you're	you are
your	belonging to you

Commonly Misspelled Words

accommodation	engineering	occurrence
auxiliary	height	parallel
basically	hydraulic	pneumatic
characteristic	hypothesis	procedure
committed	insufficient	receive
comparative	laboratory	recommend
customer	maintenance	reservoir
deteriorate	maneuverability	resonance
discrepancy	necessary	separate

Getting Started In Excel

Excel: GPA Calculator

We're going to create a grade sheet that will help you calculate your overall GPA as well as your math/science/engineering GPA.

Why you are doing this: First, most every engineer uses Excel with frequency ranging from often to constantly. Since we need to learn to use this tool anyway, we might as well do something useful. In this case, "something useful" is learning how to calculate your own GPA, and how to get a GPA specifically for your math, science, and engineering courses, so you can brag about how awesome you are on a resume one day.

1. Open a new Excel workbook.

2. Click in cell B2 and type **Semester**.

3. Click in cell C2 and type **Course #**.

4. Click in cell D2 and type **Course Name**.

5. Click in cell E2 and type **Credit Hours**.

6. Click in cell F2 and type **Grade**. Then, without doing anything else, hold down the alt key and hit enter. Then type **(A - 4, B - 3, C - 2, D - 1, F - 0)**. (On a Mac, this keyboard combination is cntrl+option+return instead of Alt+Enter.)

7. Click in cell G2 and type **Grade Points**.

8. Click in cell H2 and type **Math/Sci/Engr Class?**. Then, without doing anything else, hold down the alt key and hit enter. Then type **(yes - 1, no - 0)**.

9. Click in cell I2 and type **MSE Credit Hours**. (MSE stands for Math/Sci/Engr.)

10. Click in cell J2 and type **MSE Grade Points**.

11. Increase the width of column D to show all the words by selecting the line between the D and E labels at the top of the columns. Your mouse will turn into a line with two arrows pointing opposite directions. Click and drag to increase the size of column D.

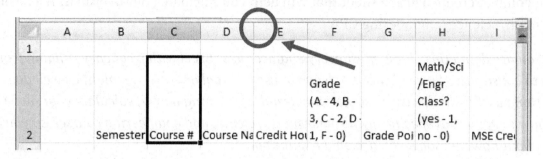

12. Do the same for columns E through J. Notice how the text we wrote in parentheses is on a separate line from the other text. That's intentional, and we did that using the `alt-enter` sequence.

13. Select cells B2 through J2 by clicking on cell B2 and dragging your mouse over to J2. Click the **B** in the ribbon above to turn the text `bold`.

14. Check and see if you need to resize any of your columns again. If so, fix any that need it.

15. Now that parenthetical stuff is getting in the way. Let's make it smaller. In cell F2, select just the text in the parentheses. (You may need to double-click inside the cell to do this.)

16. Change the font size of just that text to 9.

17. Do the same thing to the parenthetical text in cell H2. Your workbook should now look something like this.

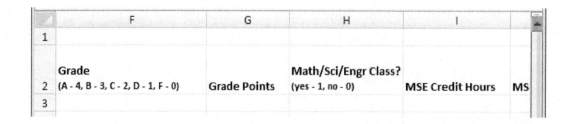

Excel: GPA Calculator

18. Now select the cells B2 through J2 and center the text by clicking the Center button in the ribbon.

19. Enter any information you have about the courses you've taken to fill out columns **_B through F only_**. Include any classes you're currently taking and what grades you think you'll get in them. A few notes:
    - *Enter 0 credit hours for any developmental classes,* even if they are in math. They do not count toward your GPA.
    - Enter the point grade (0 through 4) that you earned in the class, not the letter grade.
    - If you've retaken a class, at some schools only the most recent attempt at the class counts in your GPA. To make it easy on yourself, just don't include any times that you've taken the class in the past, just the most recent time.
    - At many schools, the second digit in a course number tells you how many credit hours the class is worth. Engr 1201 is worth 2 credit hours. Other schools may do this differently.

20. To fill in Column G, we want to use an equation. Select cell G3 and type **=E3*F3**.

21. Make sure you have clicked on cell G3. To drag the formula down, hover over the bottom right corner of that cell. Your mouse should turn into a plus sign.

22. With a plus-sign cursor, click your mouse, and drag it all the way down to the bottom row that you have filled out.

23. Check the numbers and see if they make sense.

24. Resize any columns that need resizing. Your workbook should now look something like this.

	A	B	C	D	E	F	G	H	I	J
1										
2		Semester	Course #	Course Name	Credit Hours	Grade (A - 4, B - 3, C - 2, D - 1, F - 0)	Grade Points	Math/Sci/Engr Class? (yes - 1, no - 0)	MSE Credit Hours	MSE Grade Points
3		Spring 2011	Engl 1301	English Comp 1	3	3	9			
4		Spring 2011	Math 311	Intermediate Algebra	0	4	0			
5		Spring 2011	Hist 1301	History I	3	2	6			
6		Summer 1 2011	Math 1314	College Algebra	3	3	9			
7		Summer 2 2011	Math 1316	Trigonometry	3	4	12			
8		Fall 2011	Engr 1201	Introduction to Engineering	2	4	8			
9		Fall 2011	Engr 1304	Engineering Graphics	3	4	12			
10		Fall 2011	Math 2413	Calculus I	4	4	16			
11		Fall 2011	Chem 1411	Chemistry I	4	3	12			

25. Things are going to get kind of crowded, so let's move some things around. Right-click the H column label and select Insert. The H column should have moved over to I and there should be a new blank column H.

26. Select rows 2 through 6 by clicking on row title 2 and dragging your mouse down to 6. Right-click and select `Insert`.

27. Recall that your GPA is the total number of grade points earned divided by the total number of credit hours attempted (again, developmental courses do not count). To calculate totals, Excel uses a function called `SUM`.

    First, let's total the credit hours attempted. In cell `E4`, type **Credit Hours Attempted**.

28. In cell `E5`, type **=SUM(E8:E107)**. This will sum the first 100 courses you enter into the spreadsheet. (Once you've taken more than that you should be sufficiently competent to adjust the spreadsheet to accommodate the extra classes!)

Don't worry about adjusting column widths now – we're going to hide this information later.

29. Now we need to calculate the total number of grade points earned. In cell `G4` type **Grade Points Earned**.

30. Now, rather than retyping another sum formula, let's copy the formula over from the previous cell we just messed with. Click on cell `E5`.

31. Click `Copy` from the ribbon.

32. Click on cell `G5`.

33. Click `Paste`. Now, if you double-click in cell `G5`, you'll see that the new formula you pasted now references all the data you have in that column. It adjusted all the references for you! (Pretty cool, huh?)

34. Click on cell `B2` and type **Overall GPA**.

35. Click on cell `C2` and type **=G5/E5**. Hit `enter`.

36. Format the text in `C2` to two decimal places. Right-click on cell `C2` and select `Format Cells`.

37. In the dialog box, under select the `Number` tab, the under `Category`, select `Number` again and make sure `Decimal places` is set to 2.

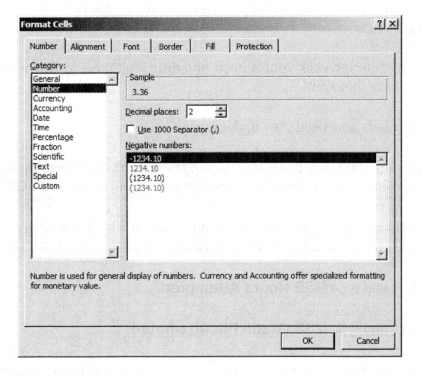

38. Click `OK`.

39. Make the text in cells `B2` and `C2` bold.

40. If you haven't already, `save` your workbook.

41. Now we're ready to pull out your Math/Science/Engineering GPA separately. In column `I`, starting in cell `I8`, begin entering a **1** if the associated class is a math, science, or engineering class, and **0** otherwise.

42. To calculate the number of Math/Science/Engineering credit hours attempted, we're going to multiply the number of credit hours times the 0/1 column you just filled out. That way, if it's not a Math/Science/Engineering class, it will say 0 hours; otherwise, it'll say the number of credit hours the class was.

    Your next step is in cell `J8` to enter the text **=E8*I8**.

43. Now you can copy this formula down for every row beneath. Rather than using copy and paste, however, we can drag it.

    Select cell J8 and hover over the bottom right corner of that cell. Your mouse should turn into a plus sign.

44. With a plus-sign cursor, click your mouse, and drag it all the way down to the bottom row that you have filled out.

45. Check the numbers and see if they make sense.

46. To calculate your Math/Science/Engineering grade points, we're going to multiply your grade points column times the 0/1 column. In cell K8 enter the text **=G8*I8**.

47. Now, rather than dragging this formula down, hover over K8 until you see the plus sign, then double-click your mouse. Your formula should automatically fill all the way down.

48. Click in cell J4 and type **MSE Hours Attempted**.

49. Click in cell K4 and type **MSE Grade Pts Attempted**.

50. Copy the formula from G5 to cells J5 and K5 using the Copy/Paste method.

51. Click in cell B3 and type **Math/Sci/Engr GPA**.

52. In cell C3, enter a formula that divides the value in K5 by the value in J5.

53. Format cell C3 to a number with two decimal places.

54. Make the text in cells B3 and C3 bold. Resize any columns if necessary.

55. Now we want to hide those rows we just used for intermediate calculations. Select rows 4 and 5. Right-click in the shaded area and select Hide.

Excel: GPA Calculator

56. Now we want to put a border around the information we calculated. Select cells `B2` through `C3` and click the border button, then click `Thick Box Border`.

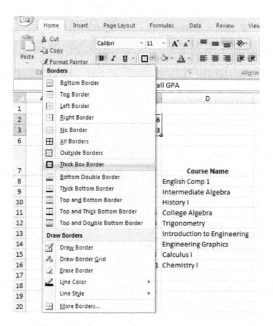

57. Celebrate! Now it will be up to you to keep this worksheet updated as you progress through your studies. `Save` your workbook and turn it in.

Excel: Grade Calculator

The goal here is to create a grade sheet that will summarize all the work done and grades earned in a particular class. The sample grading scheme that correlates with these instructions is given in the Appendix. If you are not actually using this particular grading scheme in your course and are doing this exercise for practice, make up some due dates and grades of your own.

Why you are doing this: Knowing how to calculate your class average is critical. You'll never know how well you're going if you can't keep track of what's going on. Also, your engineering professors will not be excited to hear you ask, "What do I need on the final to make an A???".

1. Open a new Excel workbook.

2. Click in cell B7 and type **Class Activities**.

3. Continue down the list, starting in cell B8, typing in the remaining grading sections for this course, as shown below.

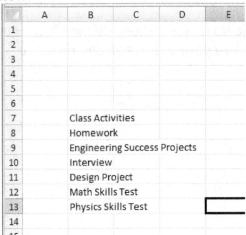

4. Select cells B7 through B13 and click Bold.

5. Autosize column B by holding the mouse in between the B and C labels (along the line – your mouse will turn into a line with two arrows pointing opposite directions) and double-clicking.

Excel: Grade Calculator

6. Click in cell C6 and type **Current Average**.

7. Click in cell D6 and type **Percent Weight**.

8. Click in cell E6 and type **Final Grade Points**.

9. Select cells C6 through E6 and select Bold, and then add a bottom border (do not select Underline).

10. Select columns C through E by clicking on the C label, and dragging your mouse over to the E label (your mouse will turn into a down arrow). Release the mouse button.

11. Autosize the columns, either by double-clicking between the C and D labels or by double-clicking between the D and E labels. Your spreadsheet should now look like this.

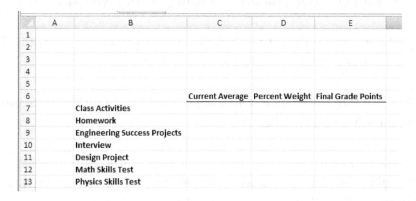

12. Click in cell D7 and type **.1**.

13. Click in cell D8 and type **.15**.

14. With cell D8 selected, hover your mouse in the right bottom corner of that cell (mouse will change to a plus-looking sign).

15. While your mouse looks like a plus sign, click your mouse, and drag your mouse until the .15 value is copied all the way down to cell D13.

16. Select cells D7 through D13 and format the text as a percent by clicking the % button.

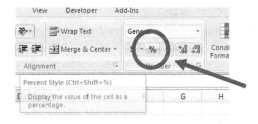

17. Now we're going to set up where we can put in grades day-to-day. Click in cell G6 and type **Class Activity**. Then type **Date** and **Grade** in cells H6 and I6, respectively.

18. Select cells G6 through I6 and make them `bold` (they may already be bold) with a `bottom border`. Also, `center` them and auto-size the columns.

19. In cell G7, type **Activity 1**.

20. Using a similar method as you did on the .15, drag the cell G7 down to cell G31. You will see that Excel automatically populates the cells with the names **Activity 2**, **Activity 3**, ... down to the maximum number of class activities you'll have this semester (indicated on your syllabus).

21. In cell H7, enter the date we did the first name quiz, which was the first class activity. Then enter every other activity date, as indicated on the syllabus.

22. Now we're going to format the dates. Select cells H7 through H31, right click in the shaded area, and click `Format Cells`....

23. Under the `Number` tab, click `Date`, and then click `3/14`. This formats the numbers to just show the month and day. Click `OK`.

24. You should have grades for the first three activities. Enter your grades. Your worksheet should look similar to that below, except we're using fake dates and grades. Your worksheet should use the actual dates and your actual grades.

	G	H	I
4			
5			
6	**Class Activity**	**Date**	**Grade**
7	Activity 1 (Name Quiz)	1/8	85
8	Activity 2 (Listening Quiz)	1/10	100
9	Activity 3 (show syllabus)	1/15	100
10	Activity 4 (success)	1/17	
11	Activity 5	1/22	
12	Activity 6	1/24	
13	Activity 7	1/29	
14	Activity 8	1/31	
15	Activity 9	2/5	
16	Activity 10	2/7	
17	Activity 11	2/12	
18	Activity 12	2/14	
19	Activity 13	2/19	

Excel: Grade Calculator

25. Now, using a similar method to steps 17 through 24 above, create a place to store your homework grades, in columns K through M. There are a few differences:
    - Instead of **Date**, write **Due Date**.
    - The due dates will be a little different, so confirm with your syllabus.
    - There may be a different number of homework assignments than class activities.
    - You may need to re-auto-size your columns.

    Your worksheet should now look similar to that below.

	K	L	M
4			
5			
6	**Homework**	**Due Date**	**Grade**
7	Homework 1	1/1	90
8	Homework 2	1/3	100
9	Homework 3	1/8	96
10	Homework 4	1/10	78
11	Homework 5	1/15	
12	Homework 6	1/17	
13	Homework 7	1/22	
14	Homework 8	1/24	
15	Homework 9	1/29	

26. Now create another space for you to store your grades for your Engineering Success Projects in columns O through Q. Recreate what you see below.

	O	P	Q
3			
4			
5			
6	**Engineering Success Projects**	**Due Date**	**Grade**
7	REU Application Project	1/1	
8	PDP & Grade Sheet		
9	Turn in #1	1/2	
10	Turn in #2	1/7	
11	Turn in #3	1/9	
12	Turn in #4	1/11	
13	Mars Rover Project		
14	Project I	1/18	
15	Project II	1/20	
16	Project III	1/22	

27. In order to make the subcategories of PDP & Grade Sheet and Mars Rover Project stand out better, we're going to indent them. To do this, first select cells O9 through O12, then press the Ctrl key on your keyboard. While holding down Ctrl, select also Cells O14 and O16. (If you make a mistake, you'll have to unselect everything and start over.) Your screen should look like this.

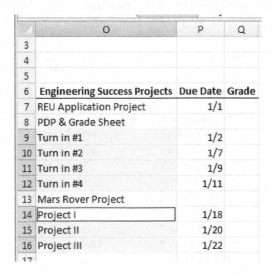

28. Now click the Increase Indent button on the Home tab.

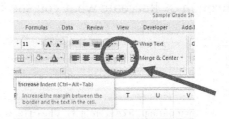

29. If you haven't already, Save your work.

30. Now let's go back to calculating your class average. Click on cell C7 and type **=average(** and then go select cells I7 through I31 and hit Enter.

31. Do the same in cell C8, but use M7 through M32 instead.

32. Do the same in cell C9, but use Q7 through Q16 instead.

33. Note now that you have a divide-by-zero error #DIV/0!. Rather than do a bunch of work to fix this, we're just going to "guess" at our grade in the first project. Go back to cell Q7 and enter what you expect to get on that project. Now the error should be gone and your spreadsheet should look similar to that below.

	A	B	C	D	E
1					
2					
3					
4					
5					
6			Current Average	Percent Weight	Final Grade Points
7		Class Activities	95	10%	
8		Homework	91	15%	
9		Engineering Success Projects	95	15%	
10		Interview		15%	
11		Design Project		15%	
12		Math Skills Test		15%	
13		Physics Skills Test		15%	

34. In cell E7, type =. Then click on cell C7. Type *. Then click on cell D7. Then hit Enter.

35. Hover your mouse over the bottom right corner of cell E7, and then double-click your mouse. The formula should autofill down to cell E13. See the image below for confirmation.

	B	C	D	E
5				
6		Current Average	Percent Weight	Final Grade Points
7	Class Activities	95	10%	9.5
8	Homework	91	15%	13.65
9	Engineering Success Projects	95	15%	14.25
10	Interview		15%	0
11	Design Project		15%	0
12	Math Skills Test		15%	0
13	Physics Skills Test		15%	0

36. Now we want to calculate your final grade in the class. In cell B2 type **Final Grade**.

37. In cell C2, type **=SUM(** and then select cells E7 through E13. Then hit enter.

38. Select cells B2 through C2 and make them bold. Then increase the font size by clicking the Increase Font Size button twice.

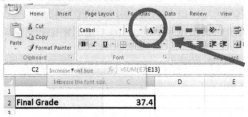

39. Select cell C2 and adjust the number of decimal places shown to 1 decimal place by either clicking Increase Decimal or Decrease Decimal.

40. Now, this is the grade you would have in this class if your Class Activities, Homework, and Engineering Success Projects averages *remained the same* and you got a **zero on everything else**.

41. The next question is, how are you doing right now in the course, given the work that you've turned in? Doing this takes a little extra math.

    First think about if all we had in this class were three tests and a final, all worth 25%.

    If you'd taken one test and earned a 100 (and the way our spreadsheet works, a zero on everything else), your average would be 100*.25 = 25. If you'd taken one test and earned a 75, your average would be 75*.25 = 18.75.

    But that's a pretty low number, as you can see with your grade. Really, you've only had a chance to earn 25% of the total grade, so it seems kind of wrong to give yourself a zero for everything else.

    Think about it if you had two test grades, a 90 and an 80. Your new average would be 90*.25 + 80*.25 = 42.5.

    Now, let's "normalize" it. Since you've only had a shot at earning 50% of the total points in the class, divide your score by .5. So, 42.5 / .5 = 85. Your current score in the class is about an 85. So assuming you do on average as well as you've done so far, you should get a B in the class.

Excel: Grade Calculator

What we're going to do is figure out how much of the grade you've had a shot at, and divide by that percent.

Since you've had Class Activities at 10%, Homework at 15%, and (we're pretending) one Success Project at 15%, you've started working toward 40% of the total grade in this class.

But how can we make Excel automatically recognize when you've done your Interview at 15%, and bump that up to 55%? (Or, 40% + another 15%.) That's what we're going to do now.

42. First, we're going to need a new column. Select column G, right click on the column, and click `Insert`. Everything should've now moved over a column.

43. Add another column between F and G.

44. Click on cell F6 and type **Is the average a number?**. (It doesn't need to be pretty since we're going to hide it later.)

45. Click on cell G6 and type **% to use for calc current grade**. (It doesn't need to be pretty since we're going to hide it later.)

46. Now, we want to tell Excel, "Only use the columns that have numbers in them." To help Excel recognize this, we're going to assume that if a cell has a number in it, then we should use it.

47. We want to find a function that tells us if a cell contains a number. To figure this out, first, select cell F7, then click the `insert function` (*fx*) button near the formula bar.

Excel: Grade Calculator

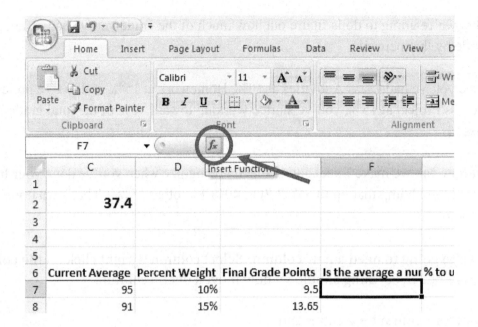

48. In the `Select a Category` box, select `Information`. We want to find information out about a particular cell.

49. Under `Select a function`, see all the different options. If you click on a function, below will have the description. Eventually, you'll see there's a function called `ISNUMBER`. The description says, "*Checks whether a value is a number, and returns TRUE or FALSE.*" That's the one we want to use. Select that function and click `OK`.

50. Now you should have a dialog box that is asking for a `Value`. What that means is it wants to know where you want it to look. Since you want to know whether or not the Current Activity category is active, select cell `C7`.

51. Before you click `OK`, see that the dialog box is already telling you what answer it's going to give. In this case, it's giving us a `TRUE`, meaning there is a number in cell `C7`.

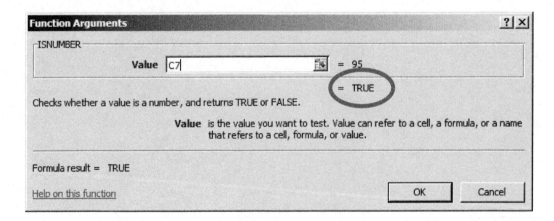

Excel: Grade Calculator

52. Click OK.

53. Now drag the formula in F7 down all the way to cell F13. Note that the cell references change, so the last formula is actually looking at whether or not you have a score entered for Physics Skills Test. Your worksheet should now look like this. Make sure the TRUE and FALSE values make sense to you.

	A	B	C	D	E	F	G
1							
2		Final Grade	37.4				
3							
4							
5							
6			Current Average	Percent Weight	Final Grade Points	Is the average a nur	% to use in calc cu
7		Class Activities	95	10%	9.5	TRUE	
8		Homework	91	15%	13.65	TRUE	
9		Engineering Success Projects	95	15%	14.25	TRUE	
10		Interview		15%	0	FALSE	
11		Design Project		15%	0	FALSE	
12		Math Skills Test		15%	0	FALSE	
13		Physics Skills Test		15%	0	FALSE	

54. Now I want to say, "If it's got a number in it, include that percentage; otherwise, don't." To figure out how to do this, first, select cell G7, then click the insert function (*fx*) button near the formula bar

55. In the Search for a function box, type **if**. We're going to look for all built-in formulas that have the word "if" in the description. Click Go.

56. One of the first items returned should be a function actually called IF. The description says, *"Checks whether a condition is met, and returns one value if TRUE, and another value if FALSE."* Select IF and click OK.

57. In the dialog box, you want to say if the value in F7 is true, then use the percent weight, otherwise use 0. To accomplish this, set the entries as follows:
    - Set logical_test to F7
    - Set Value_if_true to D7
    - Set Value_if_false to **0**

    As you can see, the formula is now giving us 10%, which is what we wanted. Click OK.

Excel: Grade Calculator

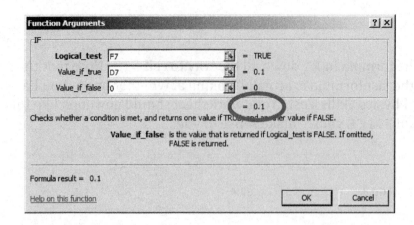

58. Now, drag that formula down to G13.

59. Sum the percentages we want to use by selecting cell G15 and typing **=SUM(** then selecting cells G7 through G13, then hitting enter. Your sum should add to 40%.

60. Save your workbook.

61. Now we're ready to get our current average in the class. In cell B4, type **Current Average**. In cell C4, create an equation that divides the value in C2 by the value in G15.

62. Now we want to change the format of cells B4 and C4 to match B2 and C2. First, select cells B2 to C2. Then click Format Painter.

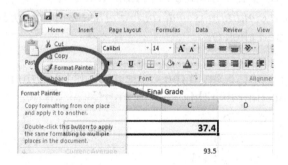

63. Now select cells B4 to C4. If you make a mistake, hit Ctrl-Z to undo and start the previous step over.

64. Yay! Now you should know your current grade in the class. Let's add a message to keep us motivated. Click in cell D4 and add an IF statement.
    - Set logical_test to C4>90
    - Set Value_if_true to **"Awesome!"**
    - Set Value_if_false to **"Keep Trying..."**

Excel: Grade Calculator

65. `Italicize` and `center` your inspirational message.

66. Now we'll add a visual indicator. Select cell C4.

67. Now click `Conditional Formatting,` click `Icon Sets,` then click `4 Ratings`.

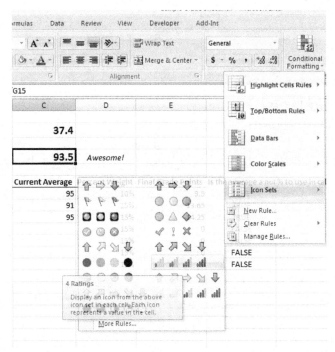

68. Now, this isn't going to work exactly as we want, so we'll want to mess with it a little. Click `Conditional Formatting` again and then click on `Manage Rules`....

69. In the dialog box, click `Icon Set`, then click `Edit Rule`....

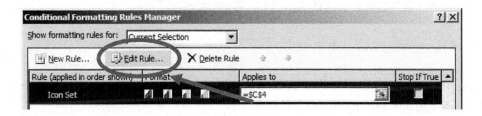

70. You want to see a different icon based on whether you have an A, B, C, or lower. You want the most bars when the value is greater than or equal to (>=) 90, one less if it's >= 80, one less if it's >= 70, and no bars if it's less than 70.

    To achieve this, match your dialog box to the one pictured below. Be sure to change `Type` to `Number`.

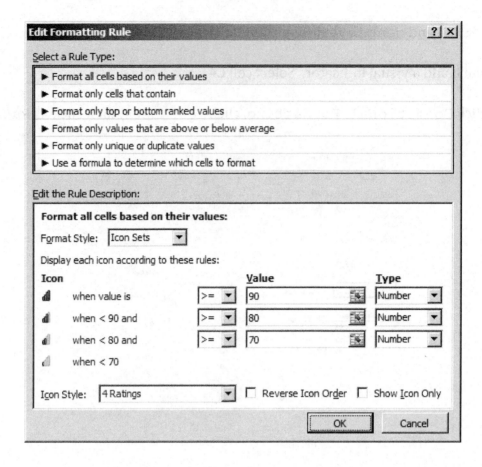

71. Click OK.

72. Click OK.

73. Now, we don't want anyone to see columns F and G anymore, so we'll hide them. Select both columns, right-click in the shaded area, and select Hide.

74. You've got it! Save your workbook.

75. Experiment by putting zeros and 100s in the various spaces in C10 through C13 and see how your grade changes.

76. Undo anything you did in the previous step and save your workbook! Enjoy!!

Excel: Graphs and Regression

Now that you have some experience with Excel, this book will start to "pull back" on the detail of instruction. No longer will you have explicit, "Click here, then click here, then click here," instructions. Instead, we will walk you through a few of the steps, and then assume that you can figure things out from there.

Again, the reason that we do this is because as an engineer, your primary job is to figure stuff out that other people won't. If you're not sure what to do, reread the information provided here, and then start poking around at different buttons to see what they do.

For this exercise, begin with the file `Graphs and Regression Source File.xlsx`. (If you do not have this file, the necessary data is contained in the Appendix.) Follow the directions in Part A, Part B, Part C, and Part D. At the end of this activity are some basic rules to follow when creating graphs.

Part A: Creating (x,y) Scatter Plots

Start with the data in the file `Graphs and Regression Source File.xlsx` in the tab labeled `Part A`.

More often than not, when you are using Excel to present data, you're not going to be creating bar graphs or pie chart. Rather, you will be displaying data over time or other measured intervals. It is important to create your chart to correctly display the data.

You can't do this with a line chart.

Time (s)	Measurement (m)
11	75.41
12	91.10
13	107.37
14	114.44
15	125.60
16	133.14
17	137.26
18	136.14
⋮	⋮
27	88.39
28	73.98
29	41.72
30	10.39

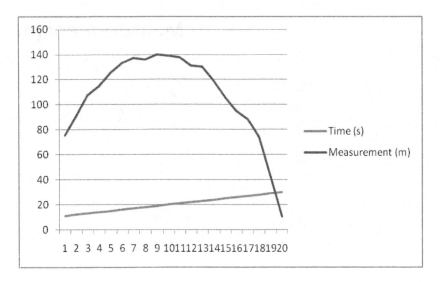

As you can see, a line chart simply plots the points as though the time increments were measured from 1 to 20. Our times go from 11 to 30. In order to plot these points correctly, we need to use an *x-y* chart, also known as a `Scatter` chart.

Select the data you wish to plot and then click `Insert` → `Scatter`, then select the plot type `Scatter with Smooth Lines and Markers`.

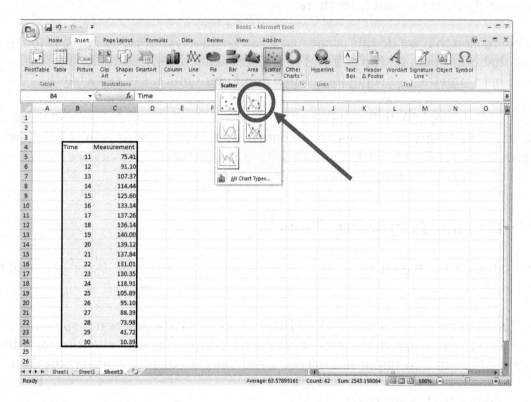

You should get a chart that looks similar to the following:

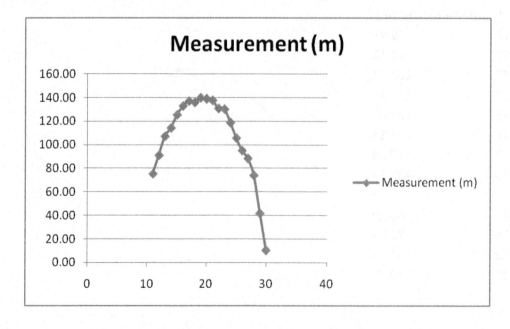

You'll want to format this with the appropriate number of significant digits and fix some other formatting. With the chart selected, click `Chart Tools`, then click `More` under `Chart Layouts`. Then select `Layout 6`.

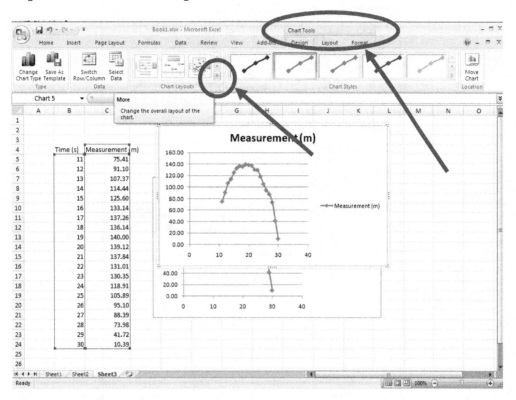

Next, delete the data label (30, 10.39) and the legend.

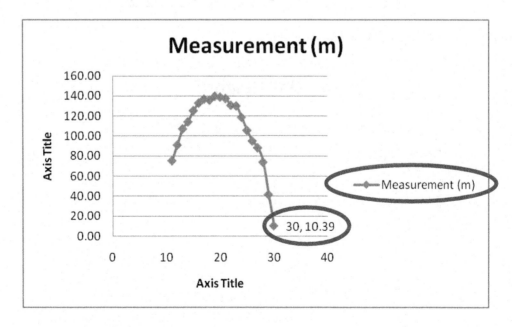

You can now modify titles for the *x* and *y* axes to say **Time (s)** and **Measurement (m)**, respectively. Modify the title to say **Ball Height**. Right click the numbers on the *y*-axis and click `Format Axis`. Then click the `Number` section and change `Decimal Places` to **0**. Click `Close`.

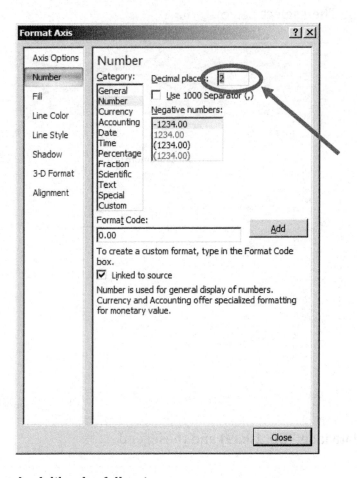

Your chart should now look like the following:

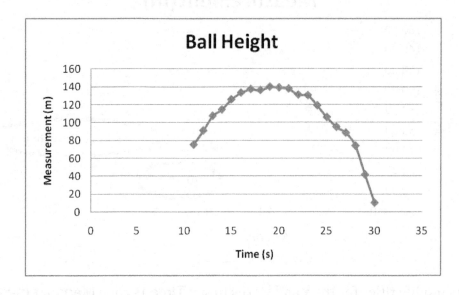

Part B: Linear Regression Plots

Start with the data in the file `Graphs and Regression Source File.xlsx` in the tab labeled `Part B`.

In this tutorial, you will use a scatter plot to create a linear regression plot. Linear regression is a statistical tool used to determine whether or not two (or more) variables are linearly related. We also sometimes call these "Best Fit" lines.

First, use the data in the Part B tab to create a scatter plot.

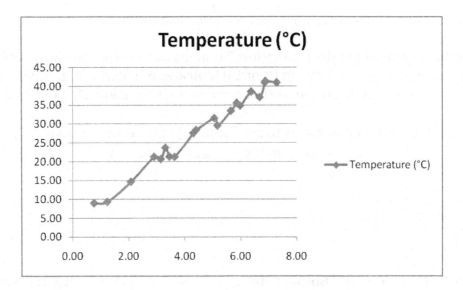

Now, we want to add the linear regression line. With the chart selected, click `Chart Tools`, then click `More` under `Chart Layouts`. Then select `Layout 9`.

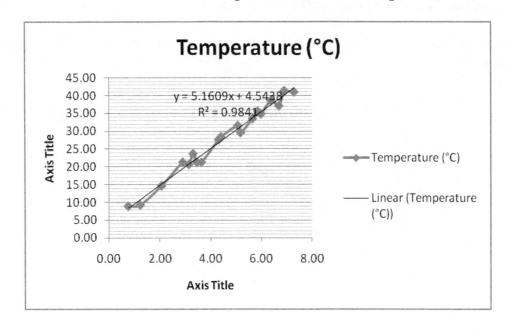

Excel: Graphs and Regression

Important!
Note that there is a new line present, and it is labeled y = 5.1609x + 4.5438 and R^2 = 0.9841. This is the linear regression (best fit) line. The equation y = ... is the line that best describes the data. If we wanted to "guess" at a point that we didn't measure, for example at time equals 2 seconds, we could substitute 2 for x in the equation.

y = 5.1609x + 4.5438
= 5.1609 · 2 + 4.5438
= 14.8656

Our best guess for the temperature at t = 2 s would be 14.8656 °C.

R^2 is the "coefficient of determination." It helps us to determine how well the regression line fits the data (an R^2 of 1.0 indicates a "perfect fit"). Since our R^2 is 0.9841, that indicates to us that our line is a very good fit of the data.

Next, modify titles for the *x* and *y* axes to say **Time (s)** and **Temperature (°C)**, respectively. Use the `Symbol` option on the `Insert` ribbon to type the degree ° symbol.

Modify the title to say **Water Temperature**. You can also delete the legend, since the regression line is labeled. Right click the numbers in the y-axis and select `Format Axis`. Then click the `Number` section and change `Decimal Places` to **0**. Click `Close`. Likewise, format the *x*-axis to display **0** decimal places.

Select the `Vertical (Value) Minor Axis Gridlines` by clicking in the graph area. Press `Delete`.

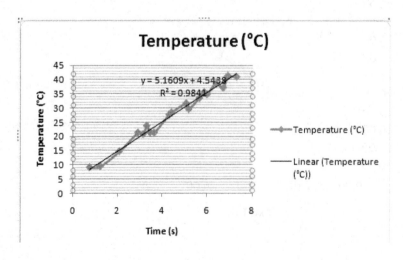

Select the Text Box with the Linear Regression information in it. Click the `Format Ribbon`. Then use `Shape Fill` and `Shape Outline` to give the text box a white background and black border. Then move the text box out of the way of the graph beneath.

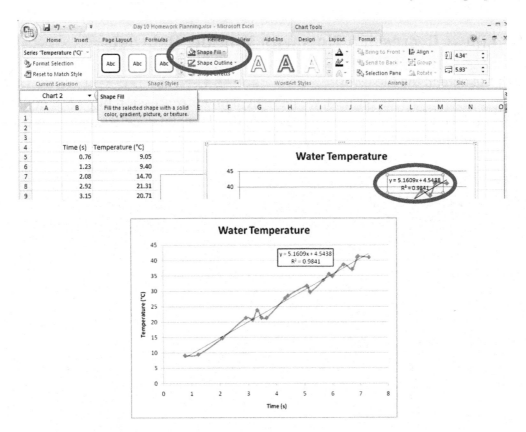

Part C: An Example of "Bad Fit" Linear Regression Plots

Start with the data in the file `Graphs and Regression Source File.xlsx` in the tab labeled `Part C` (this is the same data we used in Part A).

Create a linear regression scatter plot and format your chart to match what follows, using the methods you learned in Part A and Part B of this tutorial.

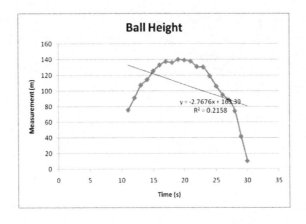

Note that the R^2 value here is 0.2158. This lets us know that a linear model of this data is a pretty bad fit. This makes sense, since the data is clearly following a parabola, and not a straight line.

Part D: Linear Regression Plots with Multiple Data Sets

Start with the data in the file `Graphs and Regression Source File.xlsx` in the tab labeled `Part D`. In this tutorial, we'll create a linear regression plot with multiple data sets, and then format the chart to best showcase the data. First, select all the data and create a scatter plot.

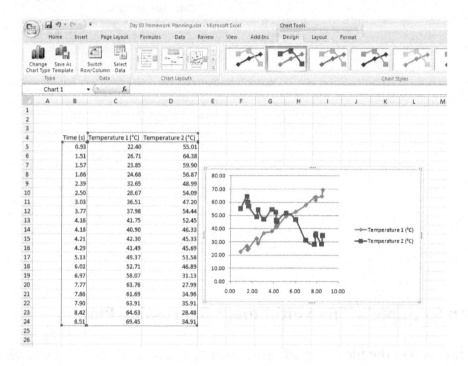

Using the methods you've learned so far, format the table to look like the example below.

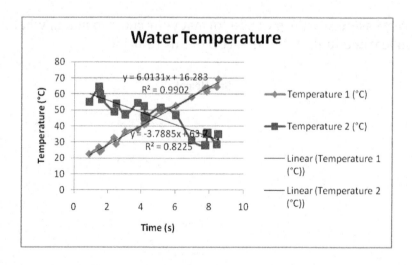

Right click the legend and select `Format Legend`. Set the `Legend Position` to `Bottom`.

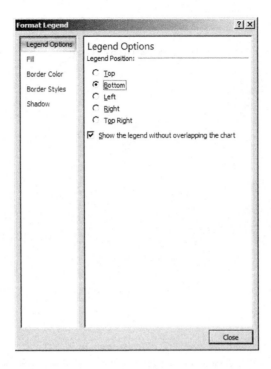

Right click the y-axis and select `Format Axis`. Next to Maximum, select Fixed, and set that value to **70**.

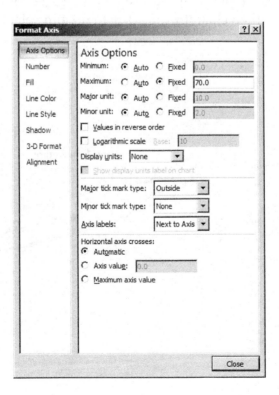

Drag the chart to make it larger. Set the Regression Information to have a white background and black border. Move the boxes so that they do not cover up any of the information.

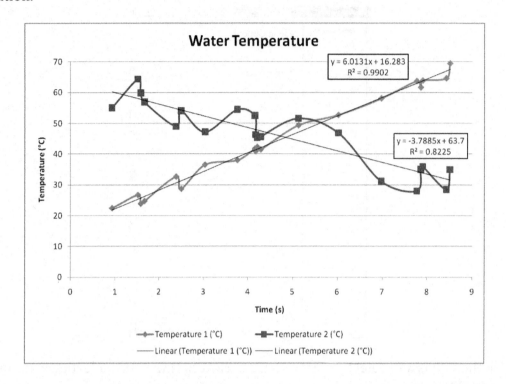

Select the line associated with the Temperature 2 linear regression. (This may take some finagling.) Select `Format`, then click `Shape Outline` and select a red color.

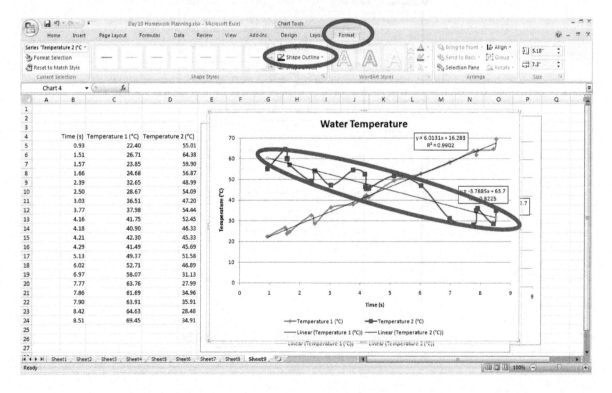

Excel: Graphs and Regression

With the line still selected, click `Shape Outline`, `Dashes`, and select `Dash`.

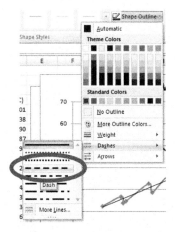

With the line still selected, click `Shape Outline`, `Weight`, and select `1½ pt`.

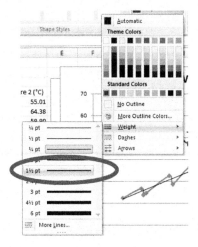

Your chart should look like this:

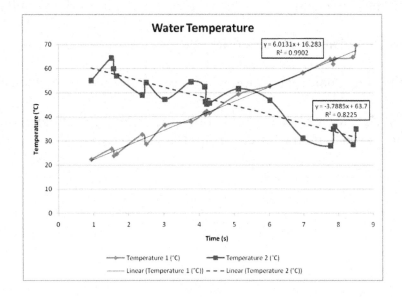

Do the same for the Temperature 1 regression line, but make it a `1½ pt, blue dashed line`.

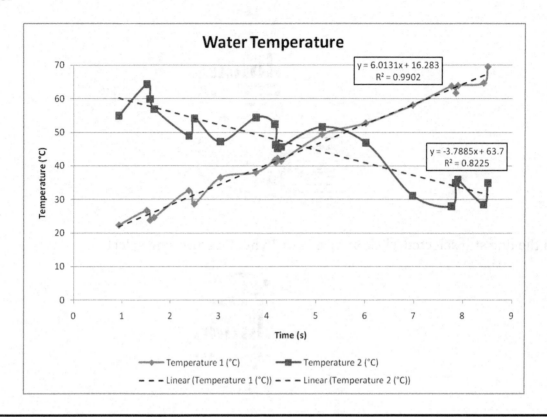

Rules for Creating Quality Graphs

Any graph made should follow these guidelines:

- Both axes should always be labeled on any graph. Not only should they be labeled with what they represent (time, distance, etc), but also with the units (m, s, lb, etc).

- Any graph should always have a title.

- If more than one quantity is graphed on the chart, you must include a legend.

- Use appropriate scales (Excel should do a pretty good job of doing this automatically) so that all your data is on the same chart

Excel: Advanced Graphing Techniques

Introduction

The purpose of this lab is to introduce probability and experimental trials, and to give you more practice using Excel to express data. Consider that you have three coins. There are 8 possible outcomes of flipping all three coins:

	1st Coin	2nd Coin	3rd Coin
Outcome 1	T	H	H
Outcome 2	T	H	T
Outcome 3	T	T	H
Outcome 4	T	T	T
Outcome 5	H	H	H
Outcome 6	H	H	T
Outcome 7	H	T	H
Outcome 8	H	T	T

To figure out the probability of flipping all heads, we count the number of outcomes that have all heads, which is just Outcome 5, and divide by the total number of outcomes, which is 8.

$1/8 = 0.125 = 12.5\%$

To figure out the probability of flipping exactly two heads, we count the number of outcomes that have exactly two heads, which are Outcome 1, Outcome 6, and Outcome 7, and divide by the total number of outcomes.

$3/8 = 0.375 = 37.5\%$

Use this information to complete the table below.

Table 1. Summary of Theoretical Probabilities

Combination	# Outcomes	Probability
All Heads	1	1/8 = 12.5%
Exactly Two Heads	3	3/8 = 37.5%
Exactly One Head		
No Heads		

You will test the predicted outcomes with actual outcomes, based on the number of trials. (The number of trials is the number of times you repeat the experiment.)

Part A: 10 Trials

How many outcomes do you expect for each combination if you flip all three coins 10 times? To get the theoretical results, multiply the probability for each combination that you calculated in the table above by the number of trials, or 10. For example, we expect to see the "All Heads" combination $0.125 \cdot 10$, or 1.25 times.

In this portion of the experiment, we'll test the theory by actually flipping all three coins, counting the number of heads, recording that value, and repeating the experiment until we have 10 trials. Complete this experiment with your lab team, and record the results in the table below. Also fill out the theoretical results for the remaining combinations. *The information in the shaded boxes should come from your Table 1.*

Table 2. Theoretical and Actual Results for 10 Trials

Combination	Theoretical Probability	Theoretical Results for 10 Trials	Actual Results
All Heads	12.5%	1.25	
Exactly Two Heads	37.5%		
Exactly One Head			
No Heads			

Part B: 50 Trials

How many outcomes do you expect for each combination if you flip all three coins 50 times? Calculate the theoretical results for 50 trials. For example, we expect to see the "All Heads" combination $0.125 \cdot 50$, or 6.25 times.

Now test the theory by actually flipping all three coins, counting the number of heads, and recording that value for 40 more trials. Use the results from your first 10 trials and add the results from the next 40 trials to get a total of 50 samples.

Complete this experiment, and record the results in the table below. Also fill out the theoretical results for the remaining combinations. *The information in the light shaded boxes should come from your Table 1. The information in the dark shaded boxes should come from your Table 2.*

Table 3 Theoretical and Actual Results for 50 Trials

Combination	Theoretical Probability	Theoretical Results for 50 Trials	Results From 10 Trials	Results From 40 More Trials	Total Results From 50 Trials
All Heads	12.5%	6.25			
Exactly Two Heads	37.5%				
Exactly One Head					
No Heads					

Part C: 100 Trials

How many outcomes do you expect for each combination if you flip all three coins 100 times? Calculate the theoretical results for 100 trials. For example, we expect to see the "All Heads" combination $0.125 \cdot 100$, or 12.5 times.

Now test the theory by actually flipping all three coins, counting the number of heads, and recording that value for 50 more trials. Use the results from your first 50 trials and add the results from the next 50 trials to get a total of 100 samples.

Complete this experiment, and record the results in the table below. Also fill out the theoretical results for the remaining combinations.

The information in the light shaded boxes should come from your Table 1. The information in the dark shaded boxes should come from your Table 3.

Combination	Theoretical Probability	Theoretical Results for 100 Trials	Results From First 50 Trials	Results From 50 More Trials	Total Results From 100 Trials
All Heads	12.5%	12.5			
Exactly Two Heads	37.5%				
Exactly One Head					
No Heads					

Now, summarize your results for all three experiments, as well as the theoretical results, expressed as a percentage.

Combination	Theoretical Results	Results From 10 Trials	Results From 50 Trials	Results From 100 Trials
All Heads	12.5%			
Exactly Two Heads				
Exactly One Head				
No Heads				

Part D: Graphing the Data

In this section, you will graph your data. Create a table in Excel that looks similar to the example below. (Obviously, your chart should have your data in it.)

	Theoretical	10 Trials	50 Trials	100 Trials
3 heads	12.5%	0	9	10
2 heads	37.5%	4	16	40
1 head	37.5%	5	21	43
0 heads	12.5%	1	4	7

Create a column chart by selecting the table you just created, and then clicking the `Insert` ribbon, then the `Column` button, and the first option under `2D Column`.

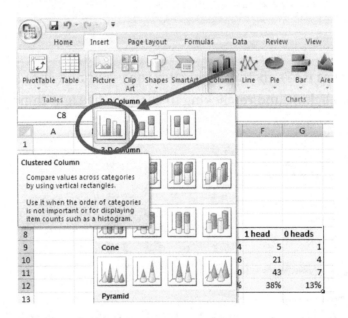

Your graph is going to be put together in a way that makes it basically impossible to compare the experiments of the three different trials.

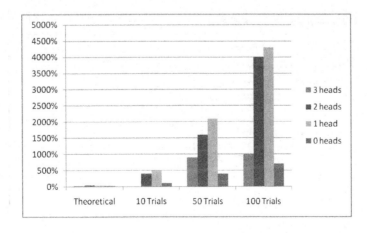

To fix this, we first need to switch the x and y axes so that we're comparing the right stuff. To do this, click the `Switch Row/Column` button.

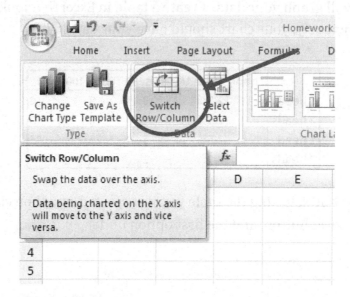

Doesn't help much, does it? To fix this, right click the *y*-axis and click `Format Axis…` (This may take some clicking and re-clicking to click on the right thing.)

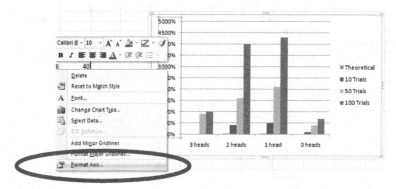

Click on `Number` on the left hand side of the dialog box, then click `Number` under Category, and then finally change the number of decimal places to **0**. Click `Close`.

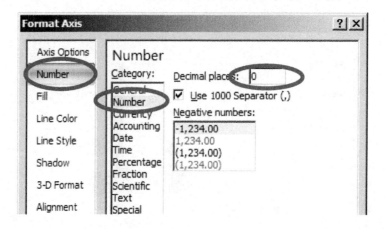

Format the layout of the chart to have a title, *x*- and *y*-axis labels, and a legend. (Remember from a previous activity that this is Chart Layout 9.) Give your chart appropriate labels. It should appear similar to the example below.

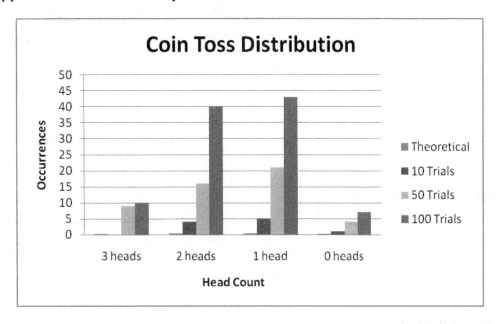

Note that it is difficult to tell exactly what this data is trying to say. This is because all of our data is on different scales. The 10 Trials data falls between 0 and 10, the 50 Trials data falls between 0 and 50, and the 100 Trials data falls between 0 and 100. The Theoretical numbers are percentages so they fall between 0 and 1.

To fix this problem, we're going to need to scale the data. Scaling the data means that we'll answer the question, "What would this data look like if we did it all the same number of times?"

The first step in scaling data is to decide what you want to scale it to. We usually scale data to nice, round numbers, like 1, 10, or 100. In this lab, we'll scale all our numbers to 100, since that is the largest data set that we have. To scale data, you take each measurement, divide by the sum of all the measurements in that experiment, and then multiply that value by your scale.

For example: For my 10 Trials data, I will take each element, divide by 10 (the total number of trials), and multiply by 100.

Category	Measurement	Sum of All Measurements	$\dfrac{\text{Measurement}}{\text{Sum of Measurements}} \times \text{Scale}$
3 heads	0	10	$\dfrac{0}{10} \cdot 100 = 0$
2 heads	4	10	$\dfrac{4}{10} \cdot 100 = 40$
1 head	5	10	$\dfrac{5}{10} \cdot 100 = 50$
0 heads	1	10	$\dfrac{1}{10} \cdot 100 = 10$

You can do this in Excel:

	Theoretical	10 Trials	50 Trials	100 Trials
3 heads	12.5%	0	9	10
2 heads	37.5%	4	16	40
1 head	37.5%	5	21	43
0 heads	12.5%	1	4	7

	Theoretical	10 Trials	50 Trials	100 Trials
3 heads		=F4/10*100		
2 heads				
1 head				
0 heads				

Scale your 10 Trials Data and your 50 Trials data using this method. To scale your theoretical percentages, multiply those percentages by 100. Note that your 100 Trials data stays exactly the same (it's already scaled to 100).

Excel: Advanced Graphing Techniques

You may need to change the format of your scaled theoretical values, by selecting `Number` instead of `Percentage`.

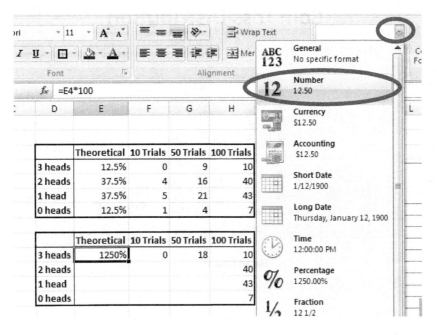

To change your chart to reflect this new data, right click your chart and click `Select Data...`. Change the data reference from your old chart to your new chart.

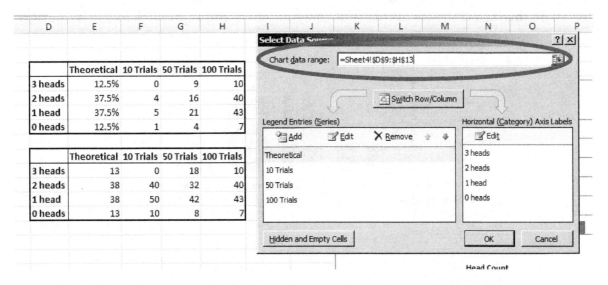

Click `OK`.

Rename your *y*-axis label **Scaled Occurrence**.

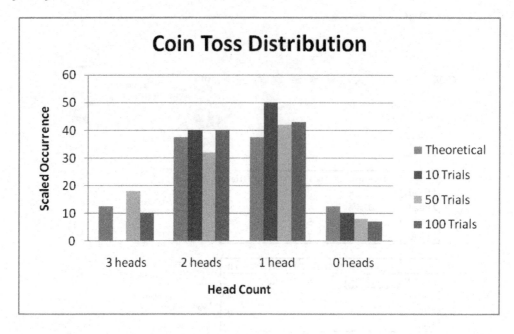

Finally, we want to see how well the actual data fits the theory. In order to emphasize this on your graph, we want to change the first data set from being another column to being a line. To do this, select the theoretical data on your chart and right-click it. Click `Change Series Chart Type...`.

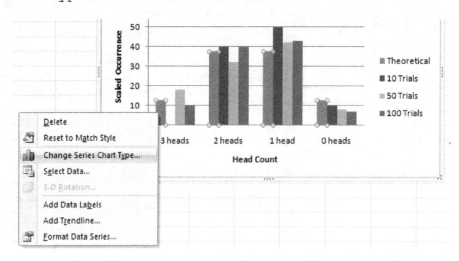

On the left side of the dialog box, select `Line`, and then select the `Line` option. Click `OK`.

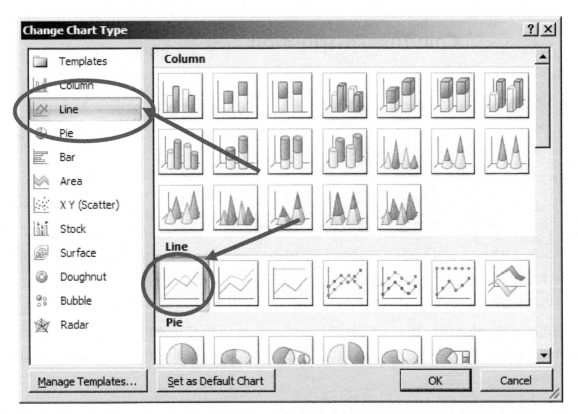

Finally, to make the line look more like a distribution, we're going to smooth it out. Right click the line you just created and then click `Format Data Series...`. In the dialog box, select `Marker Line Style`, and check the box next to `Smoothed Line`. Click `Close`.

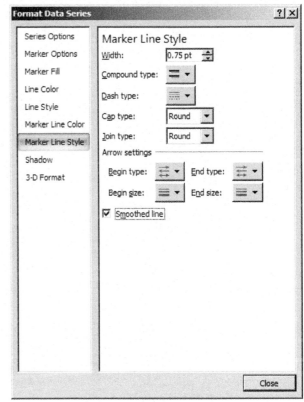

Your graph should now look similar to the following example:

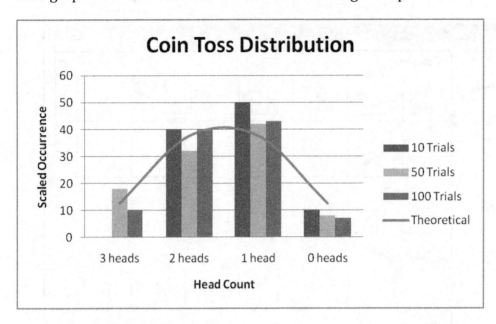

Hopefully, you will notice that the data corresponding to the higher number of trials matches the theoretical expectation better than the data from the lower number of trials.

Excel: A Case Study

For this exercise, begin with the file `Case Study.xlsm`. (If you do not have this file, the necessary data is contained in the Appendix. Be sure and save the workbook as a Macro-enabled workbook, with the extension `xlsm`.)

Introduction

One field of engineering is Systems Engineering, which includes areas such as logistics. In this exercise, you will assume the role of a systems engineer working on a team that is putting together a proposal. One section of that proposal included maintenance costs. Your team is constantly revisiting various aspects of the system to drive down maintenance costs, so every day you're getting new numbers.

One of your responsibilities is to determine the number of "spare parts" needed to maintain the system. The math required to determine the spares is straightforward.

There is a set number of operating hours that the customer expects to use the system annually. There is something else called a "Confidence Factor," which is usually set to 95%. A confidence factor is essentially, if an item on the system breaks and you go to the spares box, what's the probability that there is a spare part there.

> *Think: If you run out of paper towels, you want to be 95% sure that there are more paper towels in the pantry. That's a confidence factor.*

Confidence factors are expressed as percentages from 1% to 99%

> *Example: You can never be 100% sure that something's going to be there, so your confidence factor always has to be less than 99.999999999%. We simplify this by just limited our factors from 1% to 99%.*

Confidence factors are associated with values from the Normal Distribution table, which you will study in Statistics. For our purposes, we can just get the associated values using a function in Excel (`normsinv`). That's all you really need to know about that for this exercise.

Each item in the system has an average length of time between failures. This is called Mean Time Between Failures (MTBF). Each item also has a time required to repair, or Repair Turn Around Time (RTAT).

To calculate the spares needed, you figure out how many more items are going to fail while you're fixing the first one that broke. This is called Fails per Turn Around Time (FTAT).

The table below summarizes the calculated items and their formulas.

Calculated Item	Formula
Annual Fails (Fails per Year)	*Operating Hours / MTBF*
Fails per Day	*Annual Fails / 365*
Fails per Turn Around Time	*Fails Per Day * RTAT*
Spares Needed*	$FTAT + normsinv(Confidence\_Factor) \cdot \sqrt{FTAT}$

*We always round up Spares Needed to the next integer. The associated Excel function is `Roundup`. *Think: If you need 1.5 spare parts, then you need to buy 2. Even if you need 1.2 spare parts, you still need to buy 2.*

Part A: Setting up the Basics

For your first task, open `Case Study.xlsm`. Note that this is an "Excel Macro-Enabled Workbook," which you can tell by the `.xlsm` extension. Normal Excel workbooks have a `.xlsx` extension. When you open the workbook, you will probably get an error like the one pictured below. Be sure and click `Options...` and `Enable this Content` in the next dialog box. Then click `OK`.

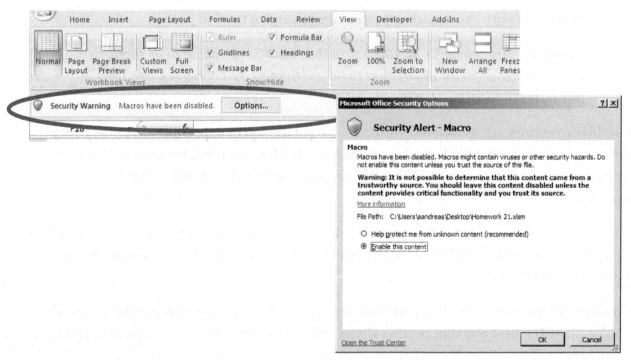

Create a new tab titled **Main**. (Recall that you create a new tab by clicking here...)

Now, recreate the spreadsheet pictured below on **Main** using the information in the narrative above. Since you have the same input, your calculated values should all be the same. To check your work, make sure your total spares cost is $661,797.

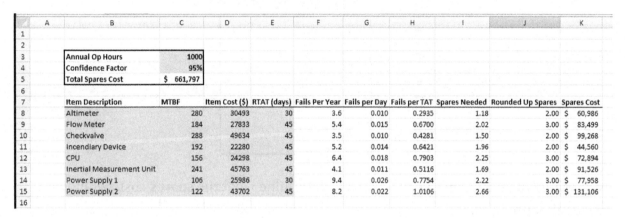

Here are some important details:

- There are four worksheets. Use the information in the worksheet titled `Data Set 1`. We'll deal with the ones called Data Sets 2 – 4 later.
- Anything highlighted in the image above is input, everything else is output. **That means that anything that is not highlighted needs to be calculated by an equation.**
- Nothing should be rounded in the spreadsheet except for the "Rounded Up Spares" column. Everything else should be controlled simply by displaying the appropriate number of decimal places using Excel formatting techniques.
- When calculating Fails Per Year, for example, the formula in cell `F8` would be **=C3/C8**. However, when you drag down that formula, it becomes **=C4/C9**, and so on, which is incorrect. You want to keep the numerator as `C3`.
    o Put a dollar sign **$** in front of the 3, which tells Excel, "When I move this formula around, don't change the 3."
    o With the formula in `F8` as **=C$3/C8**, try dragging down the formula through the rest of the rows. Now everything should look great! Format the cell to show one decimal place, and your column `F` should now look precisely as it does above.

Part B: Performing a Sensitivity Analysis

Your boss wants to know how the numbers will change if they decide to tweak the confidence factor. (This is called a Sensitivity Analysis, since it shows how sensitive the cost is to a change in input.) Create a new tab called **Confidence Comparison**. Next, create a table similar to that shown below.

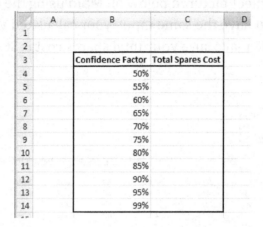

By changing the confidence factor in Main, determine what the spares costs would be at each confidence level.

> *Hint: Each time you change the confidence level, copy the spares (using* Ctrl-C*) cost off the* Main *spreadsheet, and then right-click in the cell where you want to paste the information. A pop-up menu appears. Click* Paste Special... *and select* Values. *Then click* OK.

Your numbers should match those below.

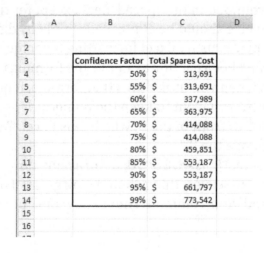

Finally, create a chart that plots the cost versus the confidence factor. You may need to change the limits on the *x*-axis. Note that you should use `Scatter with Straight Lines and Markers` instead of `Scatter with Smooth Lines and Markers`. Using smooth lines will misrepresent the data. (Try it out – can you see the difference?)

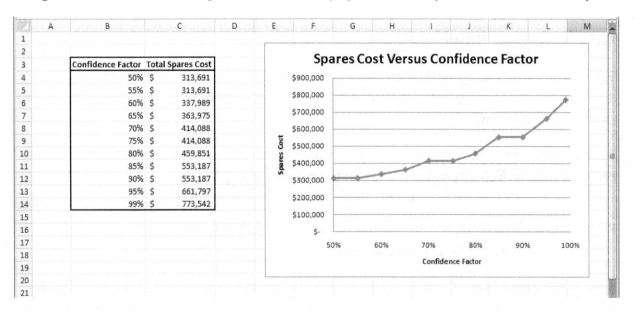

Completion of this activity is required for the following activity.

Excel: Recording Macros

This activity requires the completion of a previous activity, Excel: A Case Study.

Part A: Recording a Macro

The numbers are changing. It seems like every afternoon the reliability specialists have different MTBFs. You've now got three different sets of MTBFs, and your boss wants to see charts on all of them.

So that you don't have to do everything over and over, we'll do it one more time, but we'll record a macro. You can watch a video of the following steps at:
http://media01.mclennan.edu/vbp/aandreas/Engr_1201/CreateConfidenceComparisonMacro.wmv

First, copy the information from the tab `Data Set 2` using the `Paste Special...  Values` option into the `Main` spreadsheet.

Now, we're going to start recording the macro. In the `View` ribbon, click the arrow under `Macros` and select `Record Macro...` Name the macro **Confidence_Comparison** and in the shortcut key box type **q**. Click OK.

Starting now, anything that we do in Excel will be recorded. Any keystroke, any cell we select, *anything* will be recorded.

Note that there is now a stop button somewhere on your screen. (Exact placement may be different, depending on your settings.) Don't click it now, but eventually, when we are done recording our actions, we will stop the recording by hitting the stop button.

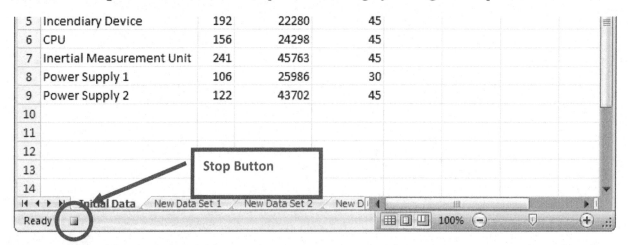

Now, we are going to go between this `Main` sheet and the `Confidence Comparison` sheet, changing the confidence value and copying the spares cost from one to the other.

But this time, all our actions will be recorded. Once you have changed all the confidence values, set the main confidence value back to **.95** and hit the stop button.

> *Hint: If you make a mistake any time when you are recording, use* `Ctrl-Z` *to undo the incorrect action. This is better than simply fixing the mistake, because using* `Ctrl-Z` *actually deletes the incorrect action in the macro. If you type a .5x instead of .5, for example, and then delete the incorrect value and retype it, all those extra steps will be in the code. That is, every time you run the macro, the computer will type a .5x, delete it, and replace it with a .5. To avoid this kind of silliness, use* `Ctrl-Z`.

You now have a new chart with all the updated values, and you have a macro that will create the table and update the graph for you in the future, any time you want it.

Part B: Using a Macro

Since it is proposal time, and the company's future is on the line, everyone is running around like crazy gibbons. Your boss now wants the confidence comparisons for all four data sets. Create a PowerPoint Presentation with four slides titled **Data Set 1**, **Data Set 2**, **Data Set 3**, and **Data Set 4**. Underneath the title you should paste a copy of the comparison graph for the corresponding data set.

In case you haven't already, save your file. In general, you always want to save a file before running a macro, in case you accidentally mess something up.

You will need to recreate the table for the first data set. Copy and paste the data set into the `Main` sheet. To run the macro, start on the `Main` sheet, click the drop-down arrow under `Macros` in the `View` ribbon and select `View Macros`. Select `Confidence_Comparison` and click `Run`.

> *Note: If you are not on the* `Main` *sheet when you run the macro, you will get all sorts of gobbedly-goop for an answer. If this happens, close the workbook without saving, and try again.*

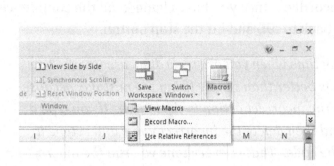 

Once you get the updated chart, you should paste it into the PowerPoint slide using `Paste Special...`, which is accessible by clicking the down arrow beneath the `Paste` button. Select `Picture (Enhanced Metafile)`. If you do not paste the chart as a picture, it will change any time the data in your Excel file changes.

After you've completed your four data slides, create an appropriate title slide and design for your presentation.

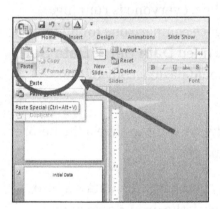 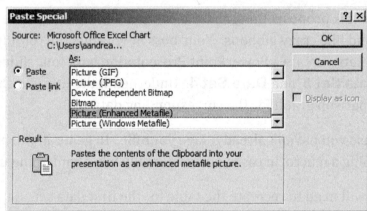

Your PowerPoint presentation should look something like this:

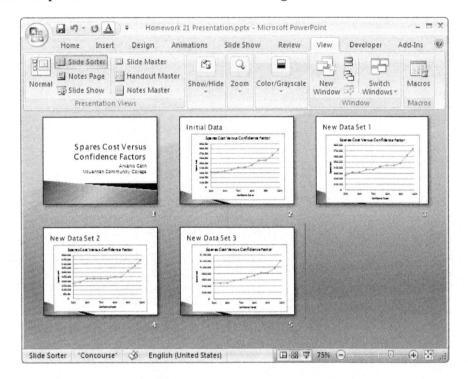

Save your presentation and close PowerPoint.

Appendices

Appendices

Appendix: Sample Grading Scheme

Grade Distribution	
Class Activities (turn in 25)	10%
Homework (turn in 26)	15%
Engineering Success Projects • REU Project (turn in 1) • PDP & Grade Sheet (turn in 4) • Mars Rover Project (turn in 3)	15%
Interview	15%
Design Project	15%
Math Skills Test	15%
Physics Skills Test	15%
Total	**100%**

Appendix: Graphs and Regression Source File.xlsx

Part A and C

Time (s)	Measurement (m)
11	75.41
12	91.10
13	107.37
14	114.44
15	125.60
16	133.14
17	137.26
18	136.14
19	140.00
20	139.12
21	137.84
22	131.01
23	130.35
24	118.91
25	105.89
26	95.10
27	88.39
28	73.98
29	41.72
30	10.39

Part B

Time (s)	Temperature (°C)
0.76	9.05
1.23	9.40
2.08	14.70
2.92	21.31
3.15	20.71
3.32	23.76
3.46	21.37
3.65	21.30
4.32	27.67
4.42	28.45
5.07	31.59
5.19	29.62
5.67	33.50
5.87	35.57
5.99	34.83
6.39	38.59
6.70	37.13
6.88	41.02
6.91	41.35
7.30	41.01

Part D

Time (s)	Temperature 1 (°C)	Temperature 2 (°C)
0.93	22.40	55.01
1.51	26.71	64.38
1.57	23.85	59.90
1.66	24.68	56.87
2.39	32.65	48.99
2.50	28.67	54.09
3.03	36.51	47.20
3.77	37.98	54.44
4.16	41.75	52.45
4.18	40.90	46.33
4.21	42.30	45.33
4.29	41.49	45.69
5.13	49.37	51.58
6.02	52.71	46.89
6.97	58.07	31.13
7.77	63.76	27.99
7.86	61.69	34.96
7.90	63.91	35.91
8.42	64.63	28.48
8.51	69.45	34.91

Appendix: A Case Study.xlsm

Data Set 1

Item Description	MTBF	Item Cost ($)	RTAT (days)
Altimeter	280	30493	30
Flow Meter	184	27833	45
Checkvalve	288	49634	45
Incendiary Device	192	22280	45
CPU	156	24298	45
Inertial Measurement Unit	241	45763	45
Power Supply 1	106	25986	30
Power Supply 2	122	43702	45

Data Set 2

Item Description	MTBF	Item Cost ($)	RTAT (days)
Altimeter	216	48416	60
Flow Meter	286	19251	30
Checkvalve	83	19381	60
Incendiary Device	295	26649	60
CPU	153	31791	30
Inertial Measurement Unit	145	14345	15
Power Supply 1	184	44847	15
Power Supply 2	136	36510	45

Data Set 3

Item Description	MTBF	Item Cost ($)	RTAT (days)
Altimeter	126	14620	15
Flow Meter	145	39427	45
Checkvalve	163	34662	15
Incendiary Device	268	10393	15
CPU	137	15834	45
Inertial Measurement Unit	298	19729	15
Power Supply 1	263	16571	45
Power Supply 2	157	24300	15

Data Set 4

Item Description	MTBF	Item Cost ($)	RTAT (days)
Altimeter	61	33167	30
Flow Meter	64	41398	60
Checkvalve	275	30371	30
Incendiary Device	158	37192	30
CPU	48	41859	60
Inertial Measurement Unit	295	13286	60
Power Supply 1	256	15151	60
Power Supply 2	179	49656	30